Reviews

In 17th century Scotland and Switzerland, two inventors independently and almost simultaneously produced the concept of logarithm, constructed tables of logarithms for efficient arithmetic, and thus started the computing revolution. John Napier (1550–1617) of Edinburgh, Baron of Merchiston, studied mathematics and astronomy at St. Salvator's College, St. Andrews. Jost Bürgi (1552–1632) of Lichtensteig, Toggenburg had only basic schooling in writing and arithmetic, never learned Latin, and thus could not read scientific literature. Nevertheless, he became an excellent mathematician, astronomer, and master craftsman of precise astronomical clocks and instruments. Subsequently, Henry Briggs (1561–1630) of London, England, derived an easier-to-use logarithm table used to this day.

For the 400-year celebration of the publication of the logarithm tables by Bürgi—developed already around 1600, but published in 1620—and the independent invention of the logarithm by Napier and Bürgi, Klaus Truemper has written a book that examines the work of these inventors. The author succeeds in bringing their thinking to life: How they decided on two very different ways to formulate the concept of logarithms and constructed quite different tables of logarithms that made efficient arithmetic possible, and how these results triggered the computing revolution that continues to this day.

The author first transports the reader to the start of the 17th century as both inventors begin their work. In particular, the reader looks over Bürgi's shoulders, so to speak, as he comes up with the idea of a table of logarithms, decides various aspects, and then computes the table in just a few months—an astonishing achievement. In contrast, Napier's and Briggs's tables require years of computing effort.

The author narrates the developments in simple, non-technical language that reads more like a detective novel than a book about mathematics. Indeed, the book provides a delightful and illuminating walk through that hugely important part of computing history.

—**Fritz Staudacher, author of the Bürgi biography** *Jost Bürgi, Kepler und der Kaiser*

The Daring Invention of Logarithm Tables takes a fresh look at the question "Who invented the concept of logarithm?". So far, the answers have frequently asserted that John Napier is the sole inventor, and that Jost Bürgi is *not* an independent co-inventor. Klaus Truemper's book explains the ideas of Jost Bürgi and John Napier, and the subsequent work of Henry Briggs, allowing the reader to trace their thinking. The text is easy to read and requires only elementary arithmetic as background.

The book invites the reader to know more about Jost Bürgi's role in this process. Probably around 1600 Bürgi had the ingenious idea to tabulate the function $f(n) = 1.0001^n$ for $n = 0, 1, \ldots, 23027$ to the precision of 9 digits. He likely accomplished this in only a few months, and he did it with no systematic error. In Napier's terminology—still used today—n is the logarithm of $f(n)$ to base 1.0001.

For the simplification of multiplication and division, the tables of Napier and Bürgi are completely equivalent. After a detailed analysis that includes results since Archimedes, the author rightfully concludes: Napier and Bürgi are independent co-inventors of the logarithms.

—**Jörg Waldvogel, Dept. of Mathematics, Eidgenössische Technische Hochschule Zürich, Switzerland**

Also by Klaus Truemper

Brain Science
Artificial Intelligence
Wittgenstein and Brain Science
Magic, Error, and Terror

History
The Construction of Mathematics

Technical
Logic-based Intelligent Systems
Effective Logic Computation
Matroid Theory

Edited by Ingrid and Klaus Truemper

F. Hülster *Introduction to Wittgenstein's*
Tractatus Logico-Philosophicus
(English and German edition)

F. Hülster *Berlin 1945: Surviving the Collapse*

THE DARING INVENTION
OF LOGARITHM TABLES

How Jost Bürgi, John Napier, and Henry Briggs
SIMPLIFIED ARITHMETIC AND
STARTED THE COMPUTING REVOLUTION

KLAUS TRUEMPER

Leibniz Company

Softcover published by Leibniz Company
2304 Cliffside Drive
Plano, Texas, 75023
USA

Original edition 2020
Updated edition 2023

Cover Art:
Left column: Bürgi image and table. Center: Briggs table. Right column: Napier image and table. Cover design by Ingrid Truemper.

The book is typeset in LaTeX using the Tufte-style book class, which was inspired by the work of Edward R. Tufte and Richard Feynman.

Sources and licenses for all figures are listed in the Notes section. The licenses implicitly cover the figures of the front cover since they are derived from images in Chapters 5, 10, 13, and 15.

Library of Congress Cataloging-in-Publication Data
Truemper, Klaus, 1942–

The Daring Invention of Logarithm Tables
Includes bibliographical references and subject index.
ISBN 978-0-9991402-0-8
1. Mathematics. 2. Logarithm

Contents

1   *Introduction*   1

2   *A Seemingly Simple Notation*   5
      Some Definitions   8

3   *Exponents*   11
      Exponents for Variables   11
      Exponents for Constants   13
      Archimedes   13

4   *Michael Stifel*   16

5   *Jost Bürgi*   19
      Decimal Number System   20
      Notation for Decimal Numbers   20
      Investigating Bürgi's Work on Logarithms   21

6   *Bürgi's Construction*   23
      An Ingenious Base   24
      Precise Form of the Base   24
      Inherent Accuracy of the Table   25
      Estimate of Computing Effort   27
      Choice of Precision   27

Actual Construction Effort   28

Representation of 10.0   28

Bürgi Constant and Scaling   28

7   *Computation with Bürgi's Scaled Table*   30

Multiplication   31

Division   32

Computation of Powers   32

Extraction of roots   33

Interpolation   34

8   *Bürgi's Table of Logarithms*   37

Accuracy of Entries   39

How did Bürgi Develop the Table?   39

9   *Instructions for Bürgi's Table*   41

Selection of Black Numbers   44

An Imagined Explanation   46

An Important Aspect   47

One More Question   48

10   *Bürgi's Title Page*   50

Two Unfortunate Decisions   51

Kepler's Comment   52

11   *Geometric Computation*   55

Invention of Circular Slide Rule   57

Invention of Slide Rule   58

12   Design of a Circular Slide Rule   61

Spacing of Ticks   61

Construction Effort   62

Improvements   62

Bürgi's Decision   63

13   John Napier   65

Napier's Model   66

Format of Napier's Table   69

Role of Differentia Column   70

Napier's Scaled Table   71

14   Computation with Napier's Table   72

Using Napier's Scaled Table   73

Computation   74

Multiplication   75

Division, Powers, Roots   75

15   Henry Briggs   76

Briggs's Scaled Table   79

Computation   80

16   Comparison of Accuracy and Efficiency   82

Accuracy   82

Bürgi's Scaled Table   82

Napier's Scaled Table   82

Briggs's Scaled Table   83

Size of Table Entries   84

Comparison of Efficiency   84

17   Beyond Bürgi, Napier, and Briggs   86
      The Difference Engine   87

18   Models of the World   90
      An Example   91
      Types of Models   92
      Modeling Errors   93
      Sources of Errors   95
      True Facts   95

19   Who Invented Logarithms?   97
      Key Steps   98
      Stumped?   101
      Stumped Again?   102
      Deciding Who Was First   103

20   Critical Comments   105
      Bürgi's Scaled Table   105
      Napier's Scaled Table   106
      Briggs's Scaled Table   106
      Objection 1   107
      Objection 2   111

21   Conclusions   115

Notes   117

Bibliography   133

Acknowledgements   136

Index   137

1

Introduction

Starting around 50 000 BCE, early humans used pebbles, scratches, or other marks to record quantities. These devices supported addition, subtraction, multiplication, and division.[1]

Eventually, various symbols replaced these tools and simplified operations. Thus, mathematics was born.

During the last 5 000 years, mathematical concepts and models became more and more advanced, and computations became more complicated as well.

In particular, the 16th and 17th centuries produced sophisticated models for puzzling observations about the world such as the configuration of the heavenly bodies. Evaluation of these models required voluminous manual computations spanning months or even years of effort.

An ingenious new computing device then reduced that effort dramatically. Indeed, the tool was an order of magnitude more effective than anything invented before.

That device was the logarithm table, or rather several such tables in particular formats. Use of these tables compressed months of computing effort to weeks and sometimes just days.

The tables triggered the invention of computing equipment where distances and angles represented the numbers of the logarithm tables. There were three such devices: the slide rule, the circular slide rule, and the slide cylinder. These tools were produced until the middle of the 20th century.[2]

A downside of the logarithm tables was their arduous and error-prone construction. That aspect led to a sequence of groundbreaking inventions. First was a special purpose computing device in the 19th century that could compute such tables mechanically and error-free. It was called the difference engine.[3] That design led to a general-purpose computer called the analytical engine[4] and the first-ever computer program.[5]

Unfortunately, both engine designs were so complicated that the difference engine was built for the first time in the late 20th and early 21st century, while the analytical engine never was—and likely never will be—constructed.

In the 1930s—about 100 years after the invention of the analytical engine—an ingenious new approach resulted in a general-purpose computer whose first version could be built using just sheet metal.[6]

The ensuing electronic revolution created ever faster computing devices that have improved human life in almost miraculous ways.

And all this started with the logarithm tables.

――――――――――――

We have taken this warp-speed tour of mathematical and computing development to highlight the extraordinary impact of the invention of the logarithm table. Thus, it surely is worthwhile to study the who, when, where, and how of this invention.

This has been done in a number of articles and books, so you may wonder why we have written yet another book on the topic.

The existing material usually recounts in the language of modern mathematics how the logarithm table was invented.

Yes, the publications typically consider that certain concepts were not available at the time. But the interpretation of events then relies on modern mathematical concepts. That approach makes the invention of the logarithm table look like a much simpler and more natural step than it actually was.

Instead, we transport ourselves into the life of one of the inventors. We look over his shoulder, so to speak, as he thinks about and works on the problem of efficient computation.

In the process, we understand how difficult the work on logarithm tables really was: Tens of thousands of computing steps had to be performed with utter precision.

At the same time, we experience the magic of this invention as it comes together in the mind of one of its creators.

You surely have noticed that so far we haven't mentioned a single name, let alone discussed the life of any person. That's done on purpose: We learn about each player as we delve into his life.

Yes, "his" is correct here. With rare exception, women at that time were thought incapable of scientific thought. Which says something about the men of that time and nothing about the women.

There has been a major controversy involving logarithms: It concerns the question of who invented logarithms first and when. So far, a variety of conflicting answers have been given.

In the last part of this book we look at these answers and come up with an explanation how such a divergence of views is possible. We also offer our own view, as you are bound to expect.

The arguments rest on a particular interpretation of the brain's reasoning process. That insight not only explains the divergence of opinions about the priority question, but also provides a reasonable justification for our answer.

One more remark about our interpretation. We are convinced that mathematics is created, and have written a book arguing the case.[7] That conclusion is contrary to that of most mathematicians, who believe that mathematics is discovered.

If you are on that side of the fence, too, don't let terms such as "invention" bother you. After all, the decision isn't based on an objective reality but depends on the models of the world we have in our heads. Chapters 9 and 23 of *Wittgenstein and Brain Science: Understanding the World*[8] discuss this aspect in detail.

————————————

This book does not require any mathematical background beyond everyday knowledge of numbers and the elementary operations of addition, subtraction, multiplication, and division.

More complicated operations are rarely used. If so, they have been moved into the Notes section.

2

A Seemingly Simple Notation

To set the stage, let x und y be variables defined as follows, where a is a number.

$$x = a \cdot a \cdot a \cdot a \cdot \ldots a \quad \text{(the factor } a \text{ occurs } m \text{ times)}$$
$$y = a \cdot a \cdot a \cdot a \cdot \ldots a \quad \text{(the factor } a \text{ occurs } n \text{ times)}$$

Evidently we have

$$x \cdot y = a \cdot a \cdot a \cdot a \cdot \ldots a \quad \text{(the factor } a \text{ occurs } m + n \text{ times)}$$

We could have expressed this more compactly by

$$x = a^m; \quad y = a^n; \quad x \cdot y = a^{m+n}$$

The restatement is very simple, right? But the notation actually is the result of a long search for a unified treatment of products of constants as well as variables.

In 1637, René Descartes (1596–1650)— outstanding philosopher, mathematician, and scientist[9]— introduced this elegant notation in an appendix titled *La Géométrie* of the book *Discours de la méthode*.

He had the same notation for variables, so $x^n = x \cdot x \cdot x \cdot x \cdot \ldots x$, where x occurs n times. He also proposed the convention that constant values

René Descartes, after Frans Hals, 1648.[10]

in such formulas be represented by the initial letters a, b, c, ... of the alphabet, and variables by the final letters ..., x, y, z.

That informal rule is still customary today.

The modern terminology for terms such as $x = a^m$ is as follows:

m is the *exponent* of a as well as the *logarithm* of x, where a is defined to be the *base*.

Furthermore, a^m is the *antilogarithm* of m, or simply a *power* of a.

DISCOURS
DE LA METHODE
Pour bien conduire fa raifon, & chercher
la verité dans les fciences.
PLUS
LA DIOPTRIQVE.
LES METEORES.
ET
LA GEOMETRIE.
Qui font des effais de cete METHODE.

A LEYDE
De l'Imprimerie de IAN MAIRE.
clɔ Iɔ c xxxvII.
Auec Priuilege.

Descartes's *Discours de la méthode*, 1637.[11]

The above expressions $x = a^m$, $y = a^n$, and $x \cdot y = a^{m+n}$ give us a clue how multiplication of x and y can be carried out:

For the given base a, find the logarithms of x and y, getting m and n. Add m and n, then determine the antilogarithm a^{m+n} as the desired value of $x \cdot y$.

If we can easily determine the logarithms m and n of x and y as well as the antilogarithm a^{m+n} of $m + n$, then we have reduced multiplication of numbers to addition of their logarithms.[12]

Similar arguments reduce division to subtraction, computation of powers to multiplication, and extraction of roots to division.

Given these seemingly obvious relationships, one is inclined to say that the idea of logarithm and its use for arithmetic is really very simple.

And yet, there has been a major controversy about who invented that concept and when, as we shall see later.

How is that possible? Why such a fuss about that evidently simple idea?

The questions actually display a misunderstanding of the development of mathematics that is fostered by the following process:

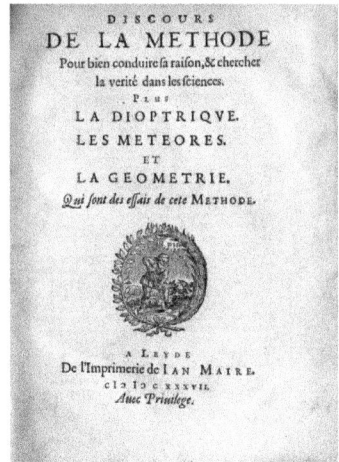

One analyzes some past achievements using modern concepts and axioms as if they were the most natural things, concludes that there is not much to those results, and then wonders why they have been declared to be significant.

In the above narration, we just engaged in that flawed process!

So let's ignore the erroneous claim that the notation $a^m \cdot a^n = a^{m+n}$ is obvious and that the concept and use of logarithms for arithmetic is self-evident.

Indeed, Jost Bürgi (1552–1632) and John Napier (1550–1617) worked hard developing that idea; as did Henry Briggs (1561–1630), who extended Napier's logarithm to the most commonly used version; or Edmund Gunter (1581–1626) and William Oughtred (1574–1660), who used Briggs's insight to create powerful computational tools.

If you want to appreciate the achievements of these inventors, imagine the following setting.

It's several centuries ago. You are an astronomer. In your work, you use the decimal numbers, each presented by an integer and a decimal fraction. An example is $3\,484\,\frac{138}{1000}$, in today's notation $3\,484.138$.

You don't know the concept of logarithms or expressions such as a^m, and you have no computing equipment except pencil and paper.

Evaluation of just one of your models of the movements of planets and stars requires hundreds of multiplications and divisions with decimal numbers. How would you do this for all the models you have postulated?

Not so easy, is it?

If you happen to live prior to the times of Bürgi and Napier, you despair since you are facing years of tedious computations.

If you live after they invented the concept of logarithm, you are lucky since their work reduces the effort to a few months.

This book narrates how Bürgi, Napier, and Briggs developed the ideas that led to efficient computations via logarithms.

Since Napier and Briggs left detailed descriptions of their ideas, it isn't too difficult to assemble the story for them. Indeed, this has already been done in great detail.[13]

The situation is different for Bürgi. He left a table of logarithms and instructions for its use, without any information how he got the key ideas or even why the instructions are correct. Hence we may only guess how Bürgi came up with his ideas about logarithms and their use.

Not to be misunderstood: There is a magnificent description of Bürgi's life and achievements,[14] a detailed treatment of the mathematical results inherent in the table,[15] an annotated English translation of the table and the instructions for its use,[16] a deep analysis of one of Bürgi's masterpieces,[17] and more.

With no criticism intended, one might say that these works tend to look at Bürgi's achievements from a modern mathematical viewpoint.

Here we try to recreate Bürgi's *thoughts* as he invents the table and works out details, using necessarily just the mathematical concepts available to him.

Before we set out on our journey, let's review the terminology connected with logarithms.

Some Definitions

Suppose $x = a^m$. As stated earlier, m is the exponent of a as well as the logarithm of x, while x is the antilogarithm of m.

Suppose we have a table with two columns. In the left column are values of x, and in the right columns the corresponding logarithms m. This arrangement is called a *table of logarithms*.

We now switch the columns, so going from left to right we obtain for a given m the corresponding x. This trivial change leads to a name change: It is now a *table of antilogarithms*.

Seems exaggerated that we change terminology just because we have switched columns, doesn't it?

Hold on, somebody is likely to object, you forgot to mention a crucial feature that is different for the two cases.

In a table of logarithms the spacing between successive x values is constant, while in a table of antilogarithms the spacing of m values is constant. These features simplify interpolation[18] of values.

Fair enough, we have to admit. The only problem with the requirement of constant spacing of x values in a table of logarithms is that it isn't satisfied by Napier's table.

Indeed, in that table neither the x values nor the m values have constant spacing when we use the table for general computation and not for special arithmetic involving sine values of trigonometry.[19]

On top, Bürgi's and Briggs's tables have admittedly small portions where spacing in both columns isn't constant. For Bürgi, this occurs in the final entries on the table, and for Briggs when we rearrange his superficially incomplete table to a smaller complete one.

So if we strictly apply the modern definitions of *table of logarithm* and *table of antilogarithm,* we are forced to conclude that neither Bürgi nor Napier nor Briggs constructed one of the two types of tables.

This doesn't make sense, particularly in light of the fact that Napier is the inventor of the word "logarithm."

Our way out is that we simply use the term *table of logarithms* for both cases and drop any requirement of constant spacing in either column. After all, our goal is to depict the glorious ideas of these inventors in terminology that focuses on common threads instead of minor differences.

––––––––––––

Before we leave the topic of naming tables, we would like to emphasize that the above classification and description is solely driven by the *use* of the tables.

There is a different viewpoint that focuses on the *construction* of the tables. When that is done, then Bürgi started with logarithms and computed antilogarithms. In simpler language, he determined powers of a given base. Thus, it is often declared that Bürgi constructed a table of antilogarithms.

On the other hand, both Napier and Briggs started with antilogarithms and computed logarithms. Hence, one may say that they created tables of logarithms.

The numerical use of each of the tables is identical: Some numbers are given, one determines their logarithms by table lookups and interpolation, manipulates these logarithms, and finally uses table lookups and interpolation to determine for the final logarithm the corresponding antilogarithm.

Seen that way, the tables of Bürgi, Napier, and Briggs have the same functionality, and we are inclined to give them the same name, which is "logarithm table."

Has this explanation succeeded in smoothing possibly-ruffled feathers? We hope so.

Let's begin the journey.

3

Exponents

The term "exponent" occurred for the first time in the 16th century, and the term "logarithm" in the 17th century. Indeed, prior to that time there was much notational confusion about exponents.

We discuss a portion of that history, starting with the notation for exponents of variables.[20]

Exponents for Variables

Diophantus of Alexandria (c. 201-214 — c. 284-298) wrote 13 books about the solution of mathematical equations. These books are referred to collectively as *Arithmetica*.

Of the 13 books, three have been lost. But some of the remaining 10 books— six Greek manuscripts and four Arabic translations—are still being reprinted more than 1700 years later.

Arithmetica relies on seemingly odd symbols for products of variables.

Arithmetica, 1621 edition, translated into Latin from Greek by Claude Gaspard Bachet de Méziriac.[21]

Using the modern notation "·" for multiplication and "=" for equality, the following statement demonstrates the confusion. Each term in the equation below is some power of the same variable.

$$\Delta^Y \cdot \Delta^Y \Delta = K^Y K$$

The riddle is solved by the following reference table, where modern notation above the line corresponds to *Arithmetica* notation below.

x	x^2	x^3	x^4	x^5	x^6
δ	Δ^Y	K^Y	$\Delta^Y \Delta$	ΔK^Y	$K^Y K$

Thus, we have the following explanation for the above equation.

$$\frac{x^2 \cdot x^4 = x^6}{\Delta^Y \cdot \Delta^Y \Delta = K^Y K}$$

Here are additional examples of exponent notation for variables.[22] Note that in each case the variable isn't explicitly shown.

Nicolas Chuquet (c. 1455 – c. 1500): The exponent of the variable is written as a superscript of the coefficient. A negative exponent is indicated by an appended ".$\tilde{m}$".

$$12^0, 12^1, 12^2, \dots \quad \text{means} \quad 12, 12x, 12x^2, \dots$$
$$12^{1.\tilde{m}} \quad \text{means} \quad 12x^{-1}$$

Pietro Antonio Cataldi (1548–1626): The exponent is appended to the coefficient as another digit.

$$53 \textit{ via } 84 \text{ fà } 407 \quad \text{means} \quad 5x^3 \cdot 8x^4 = 40x^7$$

Adrian van Roomen (1561–1615), also known as Adrianus Romanus: The exponent is encased in a box.

$$1(\overline{45}) \quad \text{means} \quad x^{45}.$$

Bürgi, whom we will discuss later in detail, used superscripted Roman numbers for exponents to clearly differentiate them from coefficients. For example, with modern notation for multiplication and equality:

$$\overset{vi}{4} \cdot \overset{iii}{3} = \overset{vi+iii}{12} = \overset{ix}{12} \quad \text{means} \quad 4x^6 \cdot 3x^3 = 12x^{6+3} = 12x^9$$

Johannes Kepler (1571–1630), who benefited greatly from Bürgi's work as we shall see later, adopted Bürgi's notation.

The situation is quite different for constants.

Exponents for Constants

In Greece of antiquity the letters of the alphabet were used to represent numbers.[23]

A (alpha) = 1, B (beta) = 2, Γ (gamma) = 3, ...

I (iota) = 10, K (kappa) = 20, Λ (lambda) = 30, ...

P (rho) = 100, Σ (sigma) = 200, T (tau) = 300, ...

$^A ⸿ = 1\,000$, $^B ⸿ = 2\,000$, $^Γ ⸿ = 3\,000$, ...; ⸿ is the archaic letter Sampi.

The largest number represented by a single letter was M (mu) = 10 000. It was called *myriad*. That term also meant "very large quantity," just as it does today.

Due to the different symbols for each power of 10, the cumbersome system fostered the notion that really large numbers could not be created. In particular, there was the belief that the number of grains of sand of the world's beaches was, if not infinite, at least so large that it could never be specified.

Archimedes (287(?)–212 BCE) proved that claim to be wrong, by showing how arbitrarily large numbers could be created.

Archimedes

Archimedes was not only the greatest mathematician of antiquity, but also an outstanding physicist, engineer, inventor, and astronomer.

He was first to compute areas above parabolas as well as ratios of volumes and surfaces for spheres and cylinders, creating a precursor of the calculus invented in the 17th century—1900 years later—by Leibniz and Newton.[24]

Of particular interest here is his proof that the number of grains of sand that would fill the entire universe has—in modern notation—fewer than $8 \cdot 10^{16}$ digits.[25]

Archimedes, by Domenico Fetti, 1620.[26]

While proving that result, he established the following landmark theorem, where the A^i are numbers. Note the text in square brackets. It paraphrases Archimedes's claim using Descartes's notation.

The lack of that notation at the time of Archimedes is the main reason the theorem is so complicated.

Theorem: If there be any number of terms of a series in continued proportion, say A_1, A_2, A_3, ..., A_m, ..., A_n, ..., A_{m+n-1}, ... of which $A_1 = 1$, $A_2 = 10$ [so that the series forms the geometrical progression 1, 10^1, 10^2, ... 10^{m-1}, ..., 10^{n-1}, ..., 10^{m+n-2}, ...], and if any two terms as A_m, A_n be taken and multiplied, the product $A_m \cdot A_n$ will be a term in the same series and will be as many terms distant from A_n, as A_m is distant from A_1; also it will be distant from A_1 by a number of terms less by one than the sum of the numbers of terms by which A_m and A_n, respectively are distant from A_1.

In modern notation, the theorem can be summarized as follows: For all $m \geq 1$ and $n \geq 1$, $10^m \cdot 10^n = 10^{m+n}$. What a simplification brought about by that notation!

400 years after Archimedes, Apollonius of Perga (240 – c. 190 BCE) defined with M and the early letters of the Greek alphabet the following numbers:

$$\overset{\alpha}{M} = 10\,000^1; \ \overset{\beta}{M} = 10\,000^2; \ \overset{\gamma}{M} = 10\,000^3; \ \overset{\delta}{M} = 10\,000^4; \dots$$

But apart from such isolated effort involving an exponent for a particular number, exponents for constants were considered unnecessary.

For example, if the case $3 \cdot 3 \cdot 3 \cdot 3$ occurred, one computed and used the resulting value 81 $(= 3 \cdot 3 \cdot 3 \cdot 3)$.

Hence, there seemingly was no need for a comprehensive notation such as 3^4.

———————

That began to change 1800 years after Archimedes, in 1544, as we see next.

4
Michael Stifel

Michael Stifel (1487–1567) had an astonishing career. He first became a priest, later sided with Martin Luther, studied mathematics, and eventually, in 1558, became the first professor of mathematics at the newly-founded University of Jena.[27]

In 1544 he published the groundbreaking book *Arithmetica Integra*. It has the following table on the reverse side of folio 249.[29]

Michael Stifel.[28]

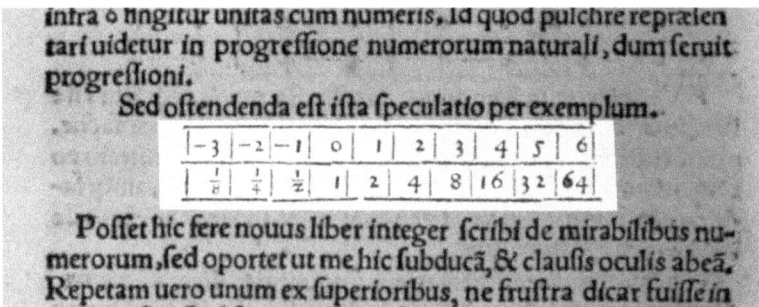

intra o fingitur unitas cum numeris, id quod pulchre repræsentari uidetur in progreſſione numerorum naturali, dum ſeruit progreſſioni.

Sed oſtendenda eſt iſta ſpeculatio per exemplum.

-3	-2	-1	0	1	2	3	4	5	6
$\frac{1}{8}$	$\frac{1}{4}$	$\frac{1}{2}$	1	2	4	8	16	32	64

Poſſet hic fere nouus liber integer ſcribi de mirabilibus numerorum, ſed oportet ut me hic ſubducā, & clauſis oculis abeā. Repetam uero unum ex ſuperioribus, ne fruſtra dicar fuiſſe in

In the top row, consecutive entries are defined by *adding* 1. Thus, they constitute an *arithmetic progression*. Stifel calls the numbers of the top row *exponents* because they are, well, "exposed."

In the bottom row, consecutive entries are produced by *multiplying* by 2. So this is a *geometric progression*. For simplicity, we call these entries just *numbers*. According to the definitions of Chapter 2, Stifel's table is a partial table of antilogarithms for base 2.

Here is Stifel's crucial observation: Suppose we *multiply* two numbers in the bottom row, getting a third number. When we *add* the exponents of those two numbers, we get the exponent of the third number.

The computational process implicitly uses the fact that, for any m and n, we have $2^m \cdot 2^n = 2^{m+n}$. This is the theorem of Archimedes for the number 2, extended to negative exponents.

As a result, if we want to multiply two numbers of the bottom row, we simply add their exponents, go to the position where that sum occurs as exponent, and find the result below that exponent.

Here is an example: Multiplication $\frac{1}{4} \cdot 8 = 2$ is reduced to addition $-2 + 3 = 1$.

$\boxed{-2}$	-1	0	1	2	$\boxed{3}$		-2	-1	0	$\boxed{1}$	2	3
$\boxed{\frac{1}{4}}$	$\frac{1}{2}$	1	2	4	$\boxed{8}$	$\Longrightarrow$	$\frac{1}{4}$	$\frac{1}{2}$	1	$\boxed{2}$	4	8

Stifel's process simplifies multiplication of numbers to addition of exponents.

Similarly, division of numbers becomes subtraction of exponents, taking of powers of a number is reduced to multiplication of its exponent by the specified power, and computation of roots of a number is accomplished by division of its exponent by the specified root.

Stifel established these relationships, but could not use them effectively: After all, the simplified computations could only be carried out if the given numbers occurred in the bottom row. But that row was far from containing all possible numbers.

The jump from Archimedes's result in Chapter 3 to Stifel's *Arithmetica Integra* in this chapter may give the erroneous impression that there was no other intervening work related to logarithms. Let's correct this.

For example, the Indian mathematician and philosopher Acharya Virasena (792–853) described the concept of *ardha cheda* in his commentary *Dhavala* of Jain mathematics.[31] Ardha cheda is the number n of times a given number x can be halved.

Acharya Virasena.[30]

If the number x is a power of 2, the ardha cheda is the binary logarithm of x. For other numbers x, it is defined by $x = 2^n \cdot y$ with y odd. Thus, it is closely related to the concept of 2-*adic order*. The latter function has the value n of ardha cheda if $n \geq 1$, but is equal to infinity if $n = 0$.[32]

Virasena covered various rules involving the concept of ardha cheda, and also described the analogous ideas for bases 3 and 4.

We have focused here on Stifel's *Arithmetica Integra* since his table is a precursor of general logarithm tables.

———————

How could Stifel's restriction to powers of 2 be removed? About 50 years after Stifel's publication, Bürgi found a way.

5
Jost Bürgi

Jost Bürgi (1552–1632) was born and grew up in Lichtensteig, Switzerland, at the time a small village. He had education only at the elementary school level and thus started his adult life with only knowledge of rudimentary mathematics and basic writing.

He never learned Latin, the customary language for scientific and mathematical publication.[33]

And yet, from such a position of disadvantage he became:

Jost Bürgi.[34]

- a master craftsman of precise clocks, mechanisms,[36] and instruments;

- a first-rate mathematician who created tools such as sine tables with unmatched precision[37] and the logarithm table discussed later;

Proportional compass.[35]

- an astronomer who supplied instruments and measurements of the heavenly bodies for Kepler and was accepted as an equal.

When Bürgi became interested in simplification of multiplication, division, computing of powers, and extraction of roots, he had already worked out the representation of decimal numbers and the related arithmetic.

Decimal Number System

Simon Stevin (ca. 1548–1620), an engineer and mathematician with wide-ranging interests, is sometimes considered the inventor of the decimal number system, since in 1585 he published the 35-page booklet *De Thiende* ("The Art of Tenths") about the decimal system in Dutch, as well as the French version *De Disme*.[40]

Mechanized celestial globe, by Jost Bürgi, 1594, Schweizerisches Landesmuseum, Zurich.[38]

But closer examination of that era reveals that there were prior and subsequent contributors, with Bürgi included in the latter group.[41]

Given the various ideas and concepts, it seems more appropriate to consider these persons, including Stevin, as co-inventors of the decimal number system and the related arithmetic.

Simon Stevin. [39]

Notation for Decimal Numbers

At the time, numerous notations were used to capture the difference between the whole number of a decimal number and the subsequent decimal fraction. Here are two examples.[42]

Stevin separated the digits of the decimal fraction by circled numbers, indicating their relative position. For example, Stevin's 348 ⓪ 5 ① 2 ② 7 ③ 9 represents the number 348.5279.

The circles were difficult to typeset. In an English translation titled *Disme: The Art of Tenths, or Decimall Arithmetike*,[43] Robert Norton (d. 1635) replaced Stevin's circles by parentheses. The notation 348 ⓪ 5 ① 2 ② 7 ③ 9 for 348.5279 thus became $348^{(0)}5^{(1)}2^{(2)}7^{(3)}9$.

Bürgi avoided such cumbersome and redundant information. He realized that he only needed to mark the point of transition from whole number to decimal fraction. He did so with a superscripted small "o" above the lowest digit of the whole number.[44] For example, his $2302\overset{o}{7}0022$ in modern notation is 230270.022. Kepler adopted Bürgi's idea, but used "(" as a separator.[45]

Napier introduced the ultimate simplification. Instead of a super- or subscripted symbol, he used a dot to separate the whole number from the decimal fraction. Thus, the decimal point was born.[46]

We examine Bürgi's work on logarithms in two steps.

Investigating Bürgi's Work on Logarithms

The first step is somewhat unusual. We look over Bürgi's shoulder while he develops his ideas, assuming—rather strangely, it may seem at the outset—that Bürgi has mathematical notation at his disposal that didn't exist during his lifetime.

Specifically, we act as if he had known Descartes's notation for exponents. Thus, Bürgi understands and uses expressions such as 10^8 or 1.0001^{1500} even though Descartes developed the notation for exponents of constants in 1637, five years after Bürgi's death in 1632.

With that assumption, we can describe Bürgi's work in modern and easy-to-understand terms. This includes a particular table of logarithms that is different from the one he actually published. At that point, we have an *understanding* of his work in the terminology of modern mathematics.

In the second step, we look at the actually published table and his accompanying instructions, and estimate how he most likely created the table and wanted it to be used. For that *interpretation* we invoke the material of the first step for illumination, but confine the arguments and conclusions to terms and concepts available at the time.

You may wonder why we proceed this way.

Since the 19th century there have been various claims and counter-claims about the interpretation of Bürgi's table and method.[47]

The flaws in some of those arguments become clear when one differentiates between understanding his work and interpreting it, as proposed here.

———————————

We start with the first step, where we imagine Bürgi's construction of the table of logarithms.

6

Bürgi's Construction

We imagine Bürgi's construction of a table of logarithms, assuming—contrary to actual events—that Bürgi has Descartes's notation available for the work.

Thus, for some base b, the table has a number of pairs of entries x and $y = b^x$.

Bürgi demands that the table can be easily constructed by manual computation, yet must be large enough to support precise multiplication and division of any numbers.

He converts these two conflicting goals into the following detailed requirements.

1. The table should only contain decimal numbers in the range 1.0–10.0, since all numbers outside this range can be readily brought into it by scaling with powers of 10 prior to the use of the table.

2. The exponents are the integers $x = 0, 1, 2, 3, \ldots$ The associated numbers $y = b^x$ should increase as well, so the selected base b must be larger than 1.0. Furthermore, the base must be such that the numbers y can be easily computed.

3. The table must have enough numbers y between 1.0 and 10.0 so that the exponents x of missing numbers can be computed with sufficient accuracy by interpolation[48] of table entries.

We continue with our fictitious story and see how Bürgi selects a base consistent with these demands.[49]

An Ingenious Base

Bürgi settles on a base of the form $1.0\ldots01$, where the exact number of 0s separating the two 1s is yet to be determined.

Regardless of the specific choice, the computation of the consecutive numbers for the table is then very easy.

For example, suppose the base is 1.0001, and we have already computed $1.0001^{3500} = 1.419\,042\,72$. The next number is

$$1.0001^{3501} = 1.0001 \cdot 1.0001^{3500} = 1.0001 \cdot 1.419\,042\,72$$

It is obtained by writing below 1.419 042 72 the same sequence of digits but shifted by four positions to the right, and then adding the two numbers. That is,

$$
\begin{aligned}
& 1.419\,042\,72 \\
+\,& 0.000\,141\,904\,272 \\
\hline
& 1.419\,184\,62
\end{aligned}
$$

Thus, Bürgi only needs elementary addition steps to construct the table. We skip here checks for rounding errors, which he certainly carries out.[50]

How does Bürgi determine the specific form for $1.0\ldots01$?

Precise Form of the Base

The closer the base $1.0\ldots01$ is to 1.0—that is, the more 0s are placed between the two 1s—the more numbers fall into the range 1.0–10.0.

Two considerations influence the choice: The inherent accuracy of the table—a concept defined next—and the workload required to compute the table.

Inherent Accuracy of the Table

Regardless of the level of precision with which the entries of the table are computed, interpolation for intermediate values introduces unavoidable errors, thus limiting accuracy.[51]

These errors occur in two ways: When an x is given and the corresponding $y = 1.0\ldots01^x$ is to be found, and when the reverse process is carried out.

The two cases are closely linked, and Bürgi most likely just investigates the error when y is to be derived for some x. We call the precision with which interpolation accomplishes this step the *inherent accuracy* of the table.

Bürgi initially explores the question of inherent accuracy using the base $b = 1.1$.

Computations with 7 digit accuracy produce in less than an hour the following table with 26 entries. The logarithm x of y is denoted by $\log_{1.1}(y)$ so that the base $b = 1.1$ is explicitly displayed.

y	$\log_{1.1}(y)$
1.000 000	0
1.100 000	1
1.210 000	2
$\cdots$	
3.138 428	12
3.452 271	13
3.797 498	14
$\cdots$	
8.954 302	23
9.849 733	24
10.000 000	24.153 approx.[52]

Next, he investigates the accuracy when he interpolates table entries to obtain y for a given logarithm value x.

Using a simple numerical test, Bürgi determines that the maximum interpolation error always occurs very close to the midpoint between adjacent logarithm values.[53] Thus he examines how well for any integer x the midpoint between 1.1^x and 1.1^{x+1} approximates $1.1^{x+0.5}$. Denote the interpolation error at that point by $D_{1.1}$.

By a straightforward process,[54] he computes $D_{1.1} = 0.0012$. Such inherent accuracy is unacceptably low. Hence the base 1.1 cannot be used.

He now computes the error D_b for the bases $b = 1.01, 1.001, 1.0001$, and so on.[55] The table below has the results.

Base b	Est. Max Error D_b	Accuracy Digits
1.1	0.0012	3
1.01	0.000 012	5
1.001	0.000 000 12	7
1.0001	0.000 000 0012	9
1.000 01	0.000 000 000 012	11

The column "Accuracy Digits" has the number n of digits after the decimal point for which interpolation is essentially accurate. We say "essentially" since we allow a 1 in the nth position of the error.

For example, for the base $b = 1.001$, we have $D_b = 0.000\,000\,12$. The first nonzero digit after the decimal point occurs in position $n = 7$ and is equal to 1. Hence, interpolation has 7 digit accuracy.

He stops with the base $b = 1.000\,01$ since the estimated max error has been reduced to the small $0.000\,000\,000\,012$. This implies an impressive 11 digit inherent accuracy.

Indeed, if the table entries by themselves have 11 digit accuracy, then computation of any y value can be accomplished with essentially identical accuracy.

How many additions are required to compute the table for that base? Bürgi investigates this question next.

Estimate of Computing Effort

We have seen that the logarithm table for the base 1.1 has $N = 24$ entries if we disregard the final one for $y = 10.0$.

What happens when the next smaller base, 1.01, is considered? Bürgi knows that $1.1 \simeq 1.01^{10}$, which means that the effort grows from $N = 24$ additions roughly by a factor 10 to $N \simeq 240$ additions.

Indeed, each time he proceeds to the next smaller base, the effort grows by the factor 10.

The base 1.001 then has a still small $N \simeq 2\,400$, while 1.0001 has an intimidating $N \simeq 24\,000$, and 1.000 01 has a huge $N \simeq 240\,000$ that rules out construction or even just printing of the table.[56]

Accordingly, he abandons the base 1.000 01 and moves to the next larger value, 1.0001. It has interpolation accuracy of 9 digits.

Choice of Precision

To utilize the 9 digit precision of interpolation, the table entries themselves must be computed to that level of accuracy.[57]

This requires a higher level of precision during the computations to prevent the buildup of roundoff errors.[58]

Evidently the choice of 9 digits is optimal in the sense that, had he selected a higher level of precision, the overall accuracy would still be limited to 9 digits due to the interpolation effect.

On the other hand, had he chosen a lower level of precision, the accuracy would have been less than 9 digits.[59]

Did Bürgi carry out the above computations or an equivalent version to select the optimal 9 digits?

It seems most likely. For once, the above computations are not difficult. And then, what are the odds that he randomly selected an optimal number of digits from the reasonable choices, say ranging from 6 to 12? Quite small, wouldn't you say?

Thus we prefer to think that the selection of the optimal 9 digits is another testimony to his meticulous workmanship in mathematics.

Bürgi is now ready to construct the table.

Actual Construction Effort

We do not know how long it takes Bürgi to determine the entire table, but likely it requires a few months. The total number of additions required for the construction turns out to be $N = 23\,027$.

For this value, Bürgi computes $1.0001^{23\,027} = 9.999\,997\,79$—the correctly rounded value is $9.999\,997\,80$—which is too small to represent 10.0 with 9 digit accuracy.

The next value, $1.0001^{23\,028} = 10.000\,997\,80$ is too large, so more work is needed to represent 10.0 in the table.

Representation of 10.0

Using interpolation with the values $9.999\,997\,79$ for the exponent $23\,027$ and $10.000\,997\,80$ for the exponent $23\,028$, Bürgi achieves 10.0 with $N = 23\,027.0022$.

Indeed, Bürgi lists $9.999\,999\,99$ for this N. That result matches the first 9 digits of the exact $1.0001^{23\,027.0022} = 9.999\,999\,996\ldots$ When rounded, either number becomes the desired 10.0.

Bürgi Constant and Scaling

Let's call the value **23 027.0022** the *Bürgi constant*. Bürgi's table—to be discussed in Chapter 8—uses red color for this constant as well as all other exponents. Here, we employ bold font for these numbers.

The Bürgi constant and scaling with powers of 10 are evidently linked as follows.

For any k, the value of 1.0001^n doesn't change if we multiply it with 10^k and subtract $k \cdot \mathbf{23\,027.0022}$ from the exponent n. Thus we have for any integer k the *Bürgi scaling*

$$1.0001^n = 10^k \cdot 1.0001^{n - k \cdot \mathbf{23\,027.0022}}$$

Shortly we also meet an *initial scaling* by powers of 10 that doesn't involve the Bürgi constant. That scaling simplifies the computing processes to come.

The table of logarithms based on the above method is a bit different from the actual table—described later in Chapter 8—by a scaling of values. Hence we call it *Bürgi's scaled table.*

The next chapter begins with an excerpt of the scaled table. We then see how it may be used for efficient multiplication, division, computation of powers, and extraction of roots.

7

Computation with Bürgi's Scaled Table

In this chapter we see how the scaled table supports simplified arithmetic. Bürgi most likely would have proceeded in a similar manner if he had known Descartes's notation for exponents of constants.

We begin with a partial display of the table. The format agrees with the 2-column arrangement described in Chapter 1.

The left column of the scaled table has numbers y ranging from 1.0 to 9.999 999 99. The right column has the corresponding red logarithms p defined by $y = 1.0001^p$; the bold font for the values simulates that color. The logarithm is denoted by $\log_{\text{Bürgi}}(y)$.

y	$\log_{\text{Bürgi}}(y)$
1.000 000 00	**0**
1.000 100 00	**1**
1.000 200 01	**2**
1.000 300 03	**3**
1.000 400 06	**4**
1.000 500 10	**5**
. . .	
3.743 174 34	**13 200**
3.743 548 65	**13 201**
3.743 923 01	**13 202**

· · ·	
9.995 998 80	**23 023**
9.996 998 40	**23 024**
9.997 998 10	**23 025**
9.998 997 90	**23 026**
9.999 997 79	**23 027**
· · ·	
9.999 999 89	**23 027.0021**
9.999 999 99	**23 027.0022**

For the discussion of example uses of the table, we reduce the number of digits of all numbers to unclutter the mathematical expressions. This includes use of a 5 digit *rounded Bürgi constant* **23 027** instead of the exact 9 digit version **23 027.0022**.

Multiplication

We want to multiply $x = 342.4$ and $y = 0.8157$. We initially scale these numbers so that each falls into the range 1.0–10.0.

Hence,

$$x \cdot y = 342.4 \cdot 0.8157 = 10^2 \cdot 3.424 \cdot 10^{-1} \cdot 8.157 = 10^1 \cdot 3.424 \cdot 8.157$$

We look up the logarithms for 3.424 and 8.157 in the scaled table. The values are **12 309** and **20 990**, respectively.

Thus,

$$3.424 \cdot 8.157 = 1.0001^{12\,309+20\,990} = 1.0001^{33\,299}$$

The exponent **33 299** is beyond the maximum value of the table. But we can always get an exponent satisfying that condition by subtracting the Bürgi constant, that is, by Bürgi scaling with $k = 1$. We get

$$3.424 \cdot 8.157 = 10^1 \cdot 1.0001^{33\,299-23\,027} = 10^1 \cdot 1.0001^{10\,272}$$

The scaled table supplies

$$1.0001^{10\,272} = 2.79309$$

We account for the various scaling steps to get the final result:

$$342.4 \cdot 0.8157 = 10^1 \cdot 3.424 \cdot 8.157 = 10^1 \cdot 10^1 \cdot 2.79309 = 279.309$$

The entire process can be summarized as follows. Note how the notation for exponents is essential for the compact description.

$$
\begin{aligned}
342.4 \cdot 0.8157 &= 10^1 \cdot 3.424 \cdot 8.157 && \text{(initial scaling)} \\
&= 10^1 \cdot (1.0001^{12\,309} \cdot 1.0001^{20\,990}) && \text{(scaled table)} \\
&= 10^1 \cdot 1.0001^{33\,299} && \text{(addition)} \\
&= 10^1 \cdot 10^1 \cdot 1.0001^{33\,299-23\,027} && \text{(B. scaling } k = 1) \\
&= 10^2 \cdot 1.0001^{10\,272} && \text{(simplify)} \\
&= 10^2 \cdot 2.79309 = 279.309 && \text{(scaled table)}
\end{aligned}
$$

Division

The division process is carried out like multiplication, except that the logarithm of the divisor is subtracted from that of the dividend. In the final step, at most addition of the Bürgi constant is needed.

Here is the summary of an example.

$$
\begin{aligned}
14.19/632.8 &= 10^{-1} \cdot (1.419/6.328) && \text{(initial scaling)} \\
&= 10^{-1} \cdot (1.0001^{3\,500}/1.0001^{18\,450}) && \text{(scaled table)} \\
&= 10^{-1} \cdot 1.0001^{-14\,950} && \text{(subtraction)} \\
&= 10^{-1} \cdot 10^{-1}1.0001^{-14\,950+23\,027} && \text{(B. scaling } k = -1) \\
&= 10^{-2} \cdot 1.0001^{8\,077} && \text{(simplify)} \\
&= 10^{-2} \cdot 2.2426 = 0.022426 && \text{(scaled table)}
\end{aligned}
$$

Computation of Powers

The computation of the kth power of a number, for some $k \geq 2$, is just as easily done as multiplication.

Let the initial scaling be by the factor 10^m. Suppose according to the table, the scaled number is equal to 1.0001^p. Using the table again, we obtain the value for $1.0001^{p \cdot k}$.

This may involve subtraction of some multiple of the Bürgi constant to get an exponent between 0 and the Bürgi constant. The number so found times $10^{k \cdot m}$ is the desired result.

We summarize the steps for an example.

$$
\begin{aligned}
14.25^8 &= 10^8 \cdot 1.425^8 && \text{(initial scaling)} \\
&= 10^8 \cdot (1.0001^{3542})^8 && \text{(scaled table)} \\
&= 10^8 \cdot 1.0001^{8 \cdot 3542} && \text{(multiplication)} \\
&= 10^8 \cdot 1.0001^{28\,336} && \text{(simplify)} \\
&= 10^8 \cdot 10^1 \cdot 1.0001^{28\,336-23\,027} && \text{(B. scaling } k=1) \\
&= 10^8 \cdot 10^1 \cdot 1.0001^{5\,309} && \text{(simplify)} \\
&= 1.7004 \cdot 10^9 && \text{(scaled table)}
\end{aligned}
$$

Extraction of roots

The extraction of roots is a bit more complicated. Say the k-th root is to be taken, for some $k \geq 2$.

The initial scaling is done by powers of 10^k in such a way that a number between 1 and 10^k is achieved. Suppose the scaling factor is $10^{k \cdot m}$.

Using Bürgi's scaled table and possibly Bürgi scaling, express the scaled number as 1.0001^q. Thus the desired root is $10^m \cdot 1.0001^{q/k}$.

Finally, the table supplies the value for $1.0001^{q/k}$. Due to the initial scaling, this final lookup never requires any Bürgi scaling.

Here is the summary for an example.

$$
\begin{aligned}
\sqrt[6]{4.05006 \cdot 10^{14}} &= 10^2 \cdot \sqrt[6]{4.05006 \cdot 10^2} && \text{(initial scaling)} \\
&= 10^2 \cdot \sqrt[6]{1.0001^{13\,988} \cdot 10^2} && \text{(scaled table)} \\
&= 10^2 \cdot \sqrt[6]{1.0001^{13\,988+2 \cdot 23\,027}} && \text{(B. scaling } k=-2) \\
&= 10^2 \cdot \sqrt[6]{1.0001^{60\,042}} && \text{(simplify)} \\
&= 10^2 \cdot 1.0001^{60\,042/6} && \text{(division by 6)} \\
&= 10^2 \cdot 1.0001^{10\,007} && \text{(simplify)} \\
&= 2.72005 \cdot 10^2 = 272.005 && \text{(scaled table)}
\end{aligned}
$$

In Chapter 6 we had Bürgi investigate interpolation when he decided on the base of the table and the accuracy of numbers.

Here we look at practical aspects.

Interpolation

Since interpolation of table values involves differences between successive table entries, well-constructed tables include those differences to simplify the task. For example, Kepler's *Tabulae Rudolphinae* and Briggs's *Arithmetica Logarithmica*—both to be discussed later—have this convenient feature.

The excerpt of Bürgi's scaled table listed earlier doesn't list these differences. We will see that Bürgi's actual table doesn't include such differences either, seemingly a shortcoming.

But quite the opposite is true: Both tables provide these differences implicitly, and interpolation can be easily carried out. Below, we do this for the scaled table.

Arcus Qua- drantis. *Cum diffe- rentiis.*	Sexa- gefima fcrupu- la.	Logarithmi *Cum diffe- rentiis.*
P. ′ ″	′ ″	
0. 0. 0	0. 0	Infinitum.
4·47		*Infinitum.*
0. 4·47	0. 5	657925.14
4·46		69314.72
0. 9·33	0.10	588610.42
4·46		40546.51
0.14·19	0.15	548063.91
4·47		28768.21
0.19. 6	0.20	519295.70

Table Heptacosias Logarithmorum Logisticorum of Kepler's *Tabulae Rudolphinae*. The first and third columns include differences in smaller font.[60]

Num. abfolu	Logarithmi.
2701	3,43152,45841,8745
	16,07604,9856
2702	3,43168,53446,8601
	16,07010,1271
2703	3,43184,60456,9872
	16,06415,7088
2704	3,43200,65872,6960
	16,05821,7299
2705	3,43216,72694,4259
	16,05228,1901
2706	3,43232,77922,6160
	16,04635,0891

Table of Briggs's *Arithmetica Logarithmica*. Differences are displayed with smaller font.[61]

Remember the earlier discussion demonstrating computation of one entry from the preceding one?

For example, the process adds to

$$1.0001^{3500.0} = 1.419\,042\,72$$

the number

$$1.419\,042\,72/10\,000 = 0.000\,141\,904\,272$$

to get the next entry

$$1.0001^{3501.0} = 1.419\,184\,62$$

In general, the difference between the two successive entries is obtained from the smaller number by shifting the decimal point four positions to the left.

Thus, the difference can be read off the table.

Here is an example interpolation where we want the number for a logarithm not listed in the scaled table.

Say that the logarithm is **12 155.3**.

The scaled table provides for **12 155** the value 3.371 774 72, and for the next **12 156** the value 3.372 111 90.

For the number corresponding to **12 155.3**, we shift the decimal point of the first number four positions to the left to get 0.000 337 18, where we have dropped irrelevant digits and rounded the last digit.

Then we add **0.3** · 0.000 337 18 = 0.000 101 15 to 3.371 774 72 to get the desired interpolated value 3.371 875 87.

Interpolation during reverse use of the scaled table is done analogously.

––––––––––

We have reached the end of the first step. At hand are Bürgi's scaled table displayed in 2-column format, and methods for multiplication, division, computation of powers, and extraction of roots. Each of these operations utilizes initial scaling, Bürgi scaling, and the Bürgi constant.

We move to the second step, where we look at Bürgi's actual table of logarithms, published in 1620, and the way he carried out the four mathematical operations.

As promised earlier, the description won't use mathematical concepts that didn't exist at Bürgi's time.

Thus, when we refer to the above results, we do so only for illumination and not for justification.

8

Bürgi's Table of Logarithms

Each page of Bürgi's table[62] contains for 400 logarithm values p the numbers 1.0001^p.

The logarithms are printed in red color; he calls them the *red numbers* ("rote Zahlen"). As before, bold font simulates that color here.

The computed values are printed in black; he declares them to be *black numbers* ("schwarze Zahlen").

All entries of the table except for one column of red numbers of the final page are whole numbers. The decimal point for those exceptions is indicated by $\overset{o}{}$ as described in Chapter 5. An additional exception occurs on the title page—discussed in Chapter 10—where 230270022 represents **230270.022**.

Due to these two cases, we know that Bürgi generally viewed the exponents, his red numbers, as whole numbers. They range from **10** to **230 270** in increments of 10. We see in Chapter 9 why he may have made that seemingly odd choice.

The computed values, his black numbers, are given as 9 digit whole numbers, except for the final 10 digit black number, which is 1 000 000 000.

Here is the first page of Bürgi's table.[63] Note the clever convention that repeated values in a column are indicated by dots.

	0	500	1000	...	3500
0	100000000	100501227	101004966	...	103561790
10	10000	11277	15067	...	72146
20	20001	21328	25168	...	82503
30	30003	31380	35271	...	92861
40	40006	41433	45374	...	103603221
50	50010	51487	55479	...	13581
⋮	⋮	⋮	⋮	⋱	⋮
450	100450991	100954479	101460489	...	104028844
460	61037	64574	70636	...	39247
470	71083	74671	80783	...	49651
480	81130	84768	90931	...	60056
490	91178	94867	101501080	...	70462
500	100501227	101004966	11230	...	80816

First page of Bürgi's table of logarithms. Bold font represents red color.[64]

The entries are interpreted as follows: The sum of the red numbers labeling a row and a column produces the black number of that row and column.

For example, the entry in column **1000** and row **450** is 101460489. This means that the red number **1 000** + **450** = **1 450** produces the black number 101 460 489.

When the red numbers are divided by 10, we get the red numbers of the scaled table.[65] Except for a few numbers on the last page of the table, the red numbers always have a **0** as rightmost digit. Hence, scaling by 10 simply means that this rightmost **0** is removed.

We obtain the black numbers of the scaled table from the above table by inserting a decimal point after the *leftmost* digit except for the number 1 000 000 000, which becomes 10.0.

Applying this rule to the above example, we see that the red **1450** and black 101460489 of Bürgi's table result in the red **145** and black 1.014 604 89 of the scaled table.

In comparison, a precise calculator produces the result $1.0001^{145} = 1.014\,604\,8994\ldots$, which matches Bürgi's number if we ignore the effect of rounding the last two digits.

It is now clear why we called the earlier displayed 2-column version *Bürgi's scaled table*: Its entries are derived from those of the original table by dividing the red numbers by 10 and inserting a decimal point into the black numbers.

Accuracy of Entries

Chapter 6 established that the inherent accuracy of the scaled table with base 1.0001 is 9 digits.

The above table shows the black numbers with exactly that many digits. Since Bürgi could easily compute the inherent accuracy with the available mathematical tools and notation—see details below—it is likely that he selected the 9 digit precision based on that consideration.

The arguments in Chapter 6 assumed that the table entries are correct. But is that the case, except maybe for a few insignificant errors?

The answer is "yes": A thorough evaluation of the accuracy of all entries of Bürgi's table concluded that it contains very few computing or typesetting errors. On top, almost all of these errors are insignificant deviations.[66]

Finally, we come to the most important question.

How did Bürgi Develop the Table?

Chapter 6 imagined Bürgi's thoughts that produced the base 1.0001, the Bürgi's constant **23 027.0022**, etc. leading up to the table. The discussion made use of Descartes's notation. But that concept did not exist at the time, so how could he have proceeded that way?

The answer is: If one goes back over those arguments, one sees that the notation is used only to simplify the representation of numbers, but not to carry out operations that Bürgi wouldn't have known.

In fact, the notation is only used to express that a certain number is multiplied with itself a certain number of times. Thus, it would be easy to rephrase the entire chapter.

For example, instead of the claim that $1.0001^{23\,028} = 10.000\,997\,80$, one might say, "1.0001 multiplied with itself 23 028 times results in 10.000 997 80." Sounds a bit cumbersome. But one could reduce the phrase to "1.0001 used 23 028 times gives 10.000 997 80" or a similar shortcut.

This is very different from advanced uses of the notation, for example to define the number $e = 2.71828\ldots$ and the natural logarithm function, which has e as base. Those concepts hadn't been defined at Bürgi's time, so Bürgi couldn't possibly have pursued thoughts involving them.[67]

Thus, our approach has followed the guideline of the introductory chapter: We have used modern notation to *understand* the table, but have relied just on concepts available at the time to *interpret* it and recreate Bürgi's likely thinking.

———————

With each copy of the table, Bürgi supplied instructions for its use. The next chapter has details. Consistent with the spirit of our investigation, we first use modern notation to understand the instructions, but then rely only on concepts available to Bürgi to clarify various steps.

9

Instructions for Bürgi's Table

At the time of publication, in 1620, Bürgi supplied the table with handwritten instructions.[68]

More than two hundred years later, in 1856, the instructions were printed as part of a paper about Bürgi.[69]

In the introduction of the instructions Bürgi discusses the fundamental link connecting the red numbers with the black ones.

Bürgi doesn't mention Stifel as source of the idea. But he states that this connection was known and had been covered by Simon Jacob (c. 1510–1564), Mauritius Zons (16th–17th c.), and quite a few others.[70]

"Wir haben in der Voredt angeregt, wie auch von etlichen Arithmeticis Simon Jacob[,] Moritius Zons und andere ist berürt worden, das was in der Geometrischen Progress oder in der Schwarzen Zahl Multipliciert dasselbige ist in der Aritmetischen Progress oder in der rothen Zahl addieren."

(In the preamble we indicated, as also touched upon by the mathematicians Simon Jacob, Moritius Zons, and quite a few others, that multiplication in the geometric progressions or black numbers is the same as addition in the arithmetic progressions or red numbers.)

It is reasonable to assume that the cited results of Jacob and Zons were based on Stifel's book *Arithmetica Integra*. Evidently, Bürgi considered the connection to be common knowledge at the time he created his table.

While discussing the example applications of the table, he defines the use of a small superscripted "o" to indicate in a decimal number the transition from whole number to fraction.[71]

"*... und werden alle Zeit bis unter die* ^{o} *ganze verstanden und die folgen der Bruch.*"

(... and the digits are always understood to represent the whole number up to below ^{o}, and the subsequent digits to represent the decimal fraction.)

Overall, the instructions seem a bit strange. They are quite verbose, yet leave some aspects for the reader to figure out.

For example, Bürgi discusses in detail multiplication of 551 192 902 with 709 153 668. Bürgi cites the corresponding red numbers **170 700**^{o} and **195 900**^{o}—the use of ^{o} here is redundant. He then adds them up to **366 600** and subtracts the Bürgi constant to get the red number **136 329**^{o}**978**. Using interpolation, he obtains from the table the corresponding black number 3 908 804 680, which, he says,[72] "*seindt die 9 ersten Ziffern des begehrten products*" (which are the first 9 digits of the desired product).

Evidently, he always looks at the final black number as a sequence of digits, and the user is to figure out the position of the decimal points in a separate consideration.

In contrast, the format for multiplication developed earlier for the scaled table accommodates the placement of decimal points. Let's do this for Bürgi's example problem, where we use the exact Bürgi constant in the Bürgi scaling so that we can reproduce Bürgi's result. For the comparison, we note again that the red numbers of the scaled table are those of Bürgi's table divided by 10.

$551\,192\,902 \cdot 709\,153\,668$

$$
\begin{aligned}
&= 10^{16} \cdot 5.511\,929\,02 \cdot 7.091\,536\,68 && \text{(initial scaling)}\\
&= 10^{16} \cdot (1.0001^{17\,070} \cdot 1.0001^{19\,590}) && \text{(scaled table)}\\
&= 10^{16} \cdot 1.0001^{36\,660} && \text{(addition)}\\
&= 10^{16} \cdot 10^{1} \cdot 1.0001^{36\,660-23\,027.0022} && \text{(B. scaling } k=1\text{)}\\
&= 10^{17} \cdot 1.0001^{13\,632.997\,8} && \text{(simplify)}\\
&= 10^{17} \cdot 3.908\,804\,680 && \text{(scaled table)}
\end{aligned}
$$

Why didn't Bürgi use this compact format and thus fully account for the decimal point?

A reasonable explanation is that our description of the computing process requires Descartes's notation for exponents of constants, which wasn't available to Bürgi. We expand upon this later.

Let's look at another example supporting this explanation. It concerns the computation of square roots.[73] We freely translate his text so that we are not distracted by unusual wording, and twice replace an erroneous "5" associated with dots and digits by "4."

"We are to extract the square root from $22\,033\,094$. First, we must add dots above the number as is customary for root extraction, and thus get $22\,033\,094$. Since we have 4 dots, the root will also have 4 digits, followed by fractions. The red number is **79 000**.

"Since the leftmost point above the black number doesn't occur on the first digit but the second one, the [Bürgi] constant must be added, and then the sum divided by 2."

He carries out these steps, but doesn't supply the final number. Via interpolation, it is $4.693\,942\,27 \cdot 10^{3}$.

Compare Bürgi's steps with the following compact description.

$$
\begin{aligned}
\sqrt[2]{2.203\,309\,4 \cdot 10^{7}} &= 10^{3} \cdot \sqrt[2]{2.203\,309\,4 \cdot 10^{1}} && \text{(initial scaling)}\\
&= 10^{3} \cdot \sqrt[2]{1.0001^{7\,900} \cdot 10^{1}} && \text{(scaled table)}\\
&= 10^{3} \cdot \sqrt[2]{1.0001^{7\,900+23\,027.0022}} && \text{(B. scaling } k=-1\text{)}\\
&= 10^{3} \cdot \sqrt[2]{1.0001^{30\,927.0022}} && \text{(simplify)}\\
&= 10^{3} \cdot 1.0001^{30\,927.0022/2} && \text{(division by 2)}\\
&= 10^{3} \cdot 1.0001^{15\,463.5011} && \text{(simplify)}
\end{aligned}
$$

$$= 4.693\,942\,27 \cdot 10^3 \qquad\qquad \text{(scaled table)}$$

Here, too, the notation for exponents is essential for the compactness of the statements.

Besides the lack of compactness of Bürgi's statements—in our opinion rooted in the lack of suitable notation—there is also a seemingly strange way in which input numbers are linked with the black numbers.

Selection of Black Numbers

The above root extraction example demonstrates the odd selection process. The input is the 8 digit number 22 033 094. But Bürgi's table has only 9 digit black numbers. Bürgi then selects via the 9 digit black number 220 330 940 the red number **79 000**.

Now suppose the input had been exactly the 9 digit black number 220 330 940. According to the earlier multiplication example he would have selected the same red number **79 000**!

This isn't the only instance where Bürgi assigns the same red number to two black numbers. An extreme case occurs when Bürgi declares that 360 000 000 has the red number **128 099789**, while 36 has the red number **128 099$\frac{78}{100}$**. Here is the statement.[74]

"Dieß ist der Schwarzen Zahl von 360 000 000 *ihr rote* **128 099789**
Es soll gleichwol verdstand werden 36 *haben ihr rothe* **128 099$\frac{78}{100}$** *"*

(For the black number 360 000 000, the red one is **128 099789**. It must be understood just the same that 36 has the red **128 099$\frac{78}{100}$**.)

According to Bürgi's interpretation of °, the first of the two red numbers corresponds to **128 099.789**, and the second one to **128 099.78**.

The difference surely is due to an error in the original manuscript or introduced during copying.

Let's just guess that Bürgi meant to specify **128 099.789** since it is the more accurate red number. Thus, he assigns to 36 and 360 000 000 the same red number **128 099789**.

Baffling? Actually, it has a straightforward explanation.

We let Bürgi himself give it in a fictitious interview, where he is shown claims and counterclaims made about his work since the 19th century and asked about his opinion.

In his response—included in a moment—he refers to Descartes's notation as well as the following function definition of Leonhard Euler (1707–1783) published in 1755, thus 123 years after Bürgi's death in 1632.

Euler's definition is close to the modern viewpoint where a function is a machine with variable values as input and function values as output.[77]

"When certain quantities depend on others in such a way that they undergo a change when the latter change, then the first are called functions of the second.

Leonhard Euler, by Jakob Emanuel Handmann, 1756.[75]

"This name has an extremely broad character; it encompasses all the ways in which one quantity can be determined in terms of others."

Here is Bürgi's imagined explanation, where we assume him not only to know Descartes's notation and the

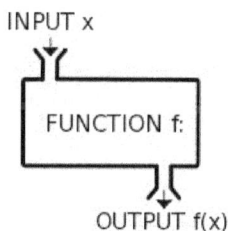

Function as a machine with input x and output $f(x)$.[76]

modern function concept, but also to have read the various interpretations of his work since the 19th century.[78]

An Imagined Explanation

"Look, these elaborate discussions about interpretation of my table and instructions miss the main point. I didn't have Descartes's notation nor Euler's concept of function. All I had was the decimal notation, of which I am one of the pioneers. Hence, any mathematical arguments employing these devices assume tools that I didn't have.

"So here is the essence. The red numbers can be changed by arbitrarily shifting the decimal point to the right or left, as long as the shift is the same for each such number. After all, the only operations we ever carry out with the red numbers are addition, subtraction, and multiplication by a constant.

"Thus, any consistent shift of the decimal point of the red numbers doesn't affect how the red numbers are connected with the black numbers.[79]

"For the selection of the black number corresponding to a given number, it doesn't matter where the decimal point occurs in that number. One only matches the 9 most significant digits of the number, or if there aren't that many digits, pads the given number with 0s to get to 9 digits.

"When the computations are completed, one figures out the position of the decimal point for the final black number.

"There is an important exception. It occurs during extraction of the kth root. Here, the position of the decimal point in the given number does matter. I account for this with properly spaced dots above the digits of the given number. The dots begin just left of the decimal point and are evenly spaced where each interval has $k - 1$ digits without dots.

"Suppose l leading digits occur to the left of the leftmost digit with dot. Then the red number is adjusted by adding l times **230270022**. [In the use of the scaled table for root extraction, the rule corresponds to the Bürgi scaling step.]

"What then really matters is the following: Each black number of the table multiplied by 1.0001 produces the successor black number and increases the red number by a fixed amount. In my table the increase is by 10. But the increase could be by any convenient constant, such as 1.

"These features are the basis for the computational effectiveness of the table.

"Given these facts, it is futile to establish a simple logarithm function that connects given numbers with the black numbers and then with the red numbers and that can be directly used for efficient computation.[80]

"Where would the exceptional handling of root extraction be fitted in? It simply cannot be done.

"This doesn't indicate a flaw in my work, but is just a symptom that thinking about computational methods was different in my days and didn't rely on the notion of functions."

With these explanations Bürgi has captured the key ideas of his ingenious invention. And when one then reads his instructions, everything is clear and simple.

An Important Aspect

Bürgi's selection of black numbers via the 9 significant digits of given numbers—simple as it is—has a beautiful side effect.

Any given number obviously results in a red number ranging from **0** to the Bürgi constant **23 027.0022**. So when two numbers are to be multiplied, the sum of their red numbers cannot exceed twice that constant.

This means that one needs to subtract from the sum at most the Bürgi constant to obtain a red number occurring in the table.

A similar conclusion applies to division, where at most the Bürgi constant needs to be added to the difference of two red numbers.

Why is this important?

If Bürgi had insisted on interpretation of the black numbers with an explicit decimal point, he would have had to add or subtract multiples of the Bürgi constant to get the correct red number for a given number.

The same difficulty would have occurred after addition or subtraction of red numbers: The resulting red number generally would have required addition or subtraction of multiples of the Bürgi constant to obtain a red number occurring in the table.

We conclude that Bürgi's matching of significant digits for the selection of black numbers isn't a simple-minded choice, but rather reflects an ingenious idea by which he avoids complicated use of the Bürgi constant.

The beneficial effect of that choice also extends to the computation of powers and the extraction of roots. Here, too, Bürgi's matching of significant digits of input numbers with black numbers eliminates unnecessary manipulation of multiples of the Bürgi constant. This is demonstrated by the earlier root extraction example.

Finally, why doesn't the use of Bürgi's scaled table—with its explicit decimal point for the black numbers and the exact matching of numbers—involve complicated multiples of the Bürgi constant?

The answer is: The initial scaling—via Descartes's notation—reduces each number to one occurring in the scaled table. Bürgi couldn't do such initial scaling since he lacked that notation.

One More Question

Before Bürgi returns to his time and place in history, we are tempted to ask him, "Why did you select the gap between the red numbers to be **10** and not **1**?"

We cannot be sure how Bürgi might have answered. Here is one guess.

"Originally I thought that all computations involving red numbers—including interpolation—could be confined to whole red numbers if the gap between successive red numbers was equal to **10**.

That was plausible for the early part of the table, since the difference between two successive numbers was small. But soon I realized that sufficient precision couldn't be achieved with that gap.

At that time, I couldn't go back and erase all those rightmost 0s of the red numbers from my notes, since I had used red ink.

"On the other hand, the computations are not affected whether the gap is **10** or **1**, or for that matter, any other number.

"So leaving the gap at **10** didn't really cause any harm."

———————

It is an enlightening exercise to line up each step of Bürgi's instructions for multiplication, division, computation of powers, and extraction of roots with the corresponding steps when the scaled table is used.

You may want to go over Bürgi's instructions and do this.[81]

In the process you will appreciate time and again the utility of Descartes's notation, and how Bürgi cleverly made do without it with his ingenious interpretation of the black numbers as significant digits.

———————

Next we look at the title page of Bürgi's table.

10

Bürgi's Title Page

The title page provides a splendid summary of the table: Every 500th red logarithm of the outer ring of the table is shown with its black number in the inner ring.

Title page of Bürgi's table.[82]

The top three lines of the title page state

"Aritmetische und Geometrische Progreß Tabulen / samt gründlichem unterricht / wie solche nützlich in allerlei Rechnungen zu gebrauchen / und verstanden werden sol"

(Arithmetic and Geometric Progression Tables / including detailed instructions / how they are useful for various calculations / and are to be interpreted)

Below the title is a handwritten comment in parentheses.

"(Dieser — nicht gedruckte — Unterricht ist im Manuscript beigefügt)" (The — not printed — instructions are added in handwritten form).

The same writer expanded the red initials "**J**" and "**B**" within the center circle to "**J***(ustus)* **B***(yrg)*".[83]

Note the small *"o"* below *"**Die ganze Rote Zahl.**"* It is a tad high above **230270022**, but belongs to that number, which thus is **230270̊022**, or in modern notation **230270.022**.

It is one of the two instances in the table where a superscripted "o" is used; the other instance occurs on the last page of the table. As we saw in Chapter 9, the instructions contain a number of such instances for both red and black numbers.

Bürgi likely computed the table of logarithms around 1600 and definitely before 1609.[84] From then he not only failed to pursue the idea of logarithm any further, but delayed publication of the table until 1620.

We look at the consequences of these unfortunate decisions.

Two Unfortunate Decisions

There are a number of possible explanations why Bürgi delayed publication of the table of logarithms for so many years.[85] We shall not explore this aspect here. But regardless of the reason, it was a most unfortunate choice.

In 1614—at least five years *after* Bürgi had computed the table, but six years *prior* to the printing of Bürgi's table—John Napier (1550–1617) published a table of logarithms that reduced arithmetic effort just as Bürgi's table did.

Napier's table was a godsend for mathematicians and scientists carrying out voluminous arithmetic operations. As a consequence, Napier became known as the inventor of logarithms.

Besides delaying publication of his table, Bürgi made another unfortunate choice: He didn't work out additional results that would have simplified or enhanced computations. Before we examine this aspect in detail, let's explore the setting of time and place.

In the years before and after 1600, Bürgi created instruments for precise measurements including a clock of extraordinary precision, contributed measurements of stars and planets, computed a very precise table of the sine values of trigonometry with a novel and highly efficient method, and last but not least, constructed the table of logarithms.

Johannes Kepler.[86]

The outstanding astronomer, mathematician, and astrologer Johannes Kepler (1571–1630) relied on these results in extensive joint work with Bürgi.

So what did Kepler think about Bürgi's decision to delay publication of the table of logarithms and not pursue that idea further?

A comment in one of Kepler's publications provides the answer.

Kepler's Comment

In 1627 Kepler published an extraordinary book that allowed computation of the positions of stars and planets when viewed from various places on earth: the *Tabulae Rudolphinae*,[87] or *Rudolphine Tables*.

The book doesn't provide the actual positions of stars and planets, but has formulas and logarithm tables with which they can be computed when viewed from any position with known latitude and longitude. To this end, the book contains the coordinates of numerous cities.

Kepler includes on page 11 of the book a comment that expresses his frustration that Bürgi had delayed publication of the table of logarithms for so

Frontispiece *Tabulae Rudolphinae*.[88]

many years and had made no effort to build upon this idea.

Page 11 of *Tabulae Rudolphinae*.[89]

"Such logistic numbers [the exponents discussed in the preceding sentence] led Justus Byrgius to the very same logarithms many years prior to the appearance of Napier's system. But he, a hesitating man and guardian of his secrets, abandoned the child at birth and didn't raise it for common benefit."[90]

Incidentally, this is incontrovertible proof that Bürgi invented the table of powers many years before Napier published his table of logarithms. We return to this point in Chapter 20.

———————————

In the next chapter we look at ideas that Bürgi surely had while working on the table of logarithms, but for some reason did not pursue. The title page provides an important clue.

11

Geometric Computation

The outer ring on the title page has the red numbers in increments of **500**. This suggests a translation of multiplication to geometric angles areas sweeping around the circle as follows.[91]

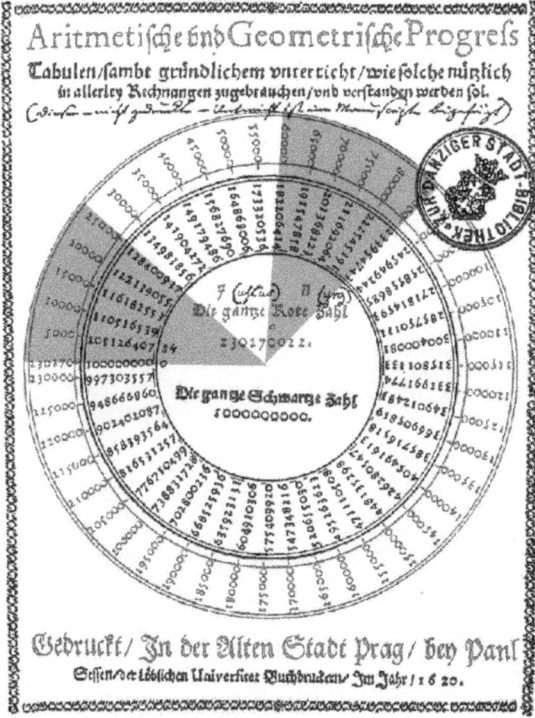

Title page with sweeping angles.[92]

Consider the left triangular area. It covers the red numbers from 0 to 2 500 and the corresponding black numbers of the inner ring. By definition of the red and black numbers, the lowest black number within the triangle multiplied by $1.0001^{2\,500}$ gives the largest black number. The second triangle has the same size, thus also sweeps a total of 2 500 red numbers, and multiplication of the smallest black number by $1.0001^{2\,500}$ produces the largest black number.

Wouldn't you assume that Bürgi had this insight? Had he not constructed precision instruments for measuring the position of stars and planets, a compass for transmitting proportions, and clocks of incredible precision before development of the table of logarithms?

So let's assume that Bürgi had that insight. Then he certainly realized that multiplication can be done graphically without use of the red numbers![93]

As a demonstration, consider $1.284 \cdot 1.822$. Define a shaded angle that covers the range 1.0–1.284. Rotate the angle clockwise until the inside of the left border touches 1.822. Then the number on the inside of the right border is the desired result $2.339 = 1.284 \cdot 1.822$, as shown below.

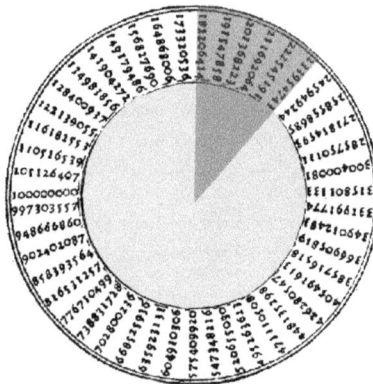

Ring of black numbers with sweeping angle.[94]

It is a small step from the ring of black numbers with a shaded angle to the circular slide rule.

However, Bürgi did not take this step.

Invention of Circular Slide Rule

The mathematician William Oughtred (1574–1660) took that step in 1632—surely without knowledge of Bürgi's table and title page—when he invented the circular slide rule.[95]

In Oughtred's design, two pointers correspond to the boundary of the shaded angle.

The pointers can be moved individually or jointly, thus allowing both definition and rotation of angles.

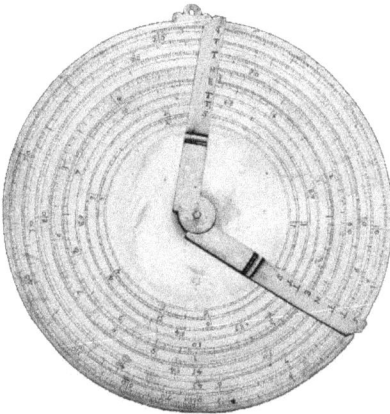

Oughtred circular slide rule.[97]

William Oughtred, by Wenceslaus Hollar.[96]

Instead of rotating a shaded angle, one could also use two nested rings of the black numbers as shown on the next page.

The reader should imagine that the rings can be rotated relative to each other to represent addition or subtraction of distances.

The photo on the right shows a modern implementation. Even though Bürgi never constructed anything of the kind, we like to call it *Bürgi's Computing Disk* since it relies on two copies of the ring of black numbers.

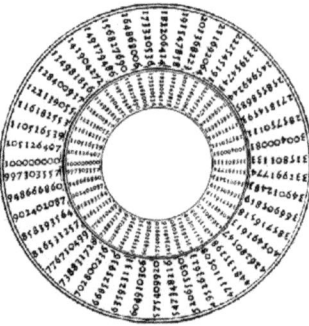

Nested rings of black numbers.[98]

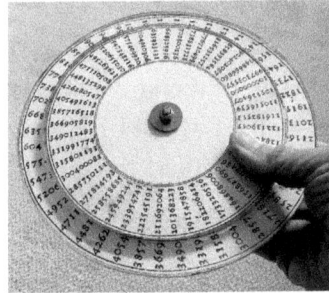

Implementation using plastic disks.[99]

As we see next, the circular slide rule was a latecomer in the use of logarithmic scales for computation.

Invention of Slide Rule

The story begins with Edmund Gunter (1581–1626). He had been mentored in mathematics by Briggs, who in 1596 had become the first professor of geometry at the recently founded Gresham College in London.

In 1619, Gunter joined that very same college as professor of geometry and thus became a colleague of Briggs. He worked in mathematics, geometry, and astronomy.

In 1620—thus one year after joining Gresham College and twelve years before Oughtred invented the circular slide rule—Gunter created the logarithmic scale base 10. It became known as *Gunter's Line*. Gunter surely obtained the data for the specification of that line from the table of logarithms base 10 constructed by Briggs. We cover that table in a later chapter.

Gunter marked his line on a ruler.

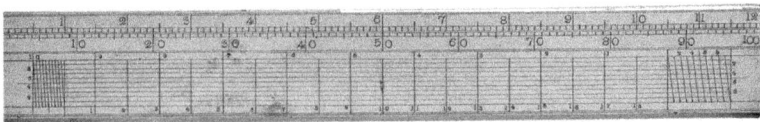

Gunter's ruler.[100]

He then declared that multiplication could be done as follows.

Say one wants to compute 1.284 · 1.822. Measure with a non-collapsing compass the distance from the 1.0 of the scale to 1.284. Next, set one leg of the compass on 1.822 and the second leg to the right on the scale. That second leg is at the desired solution 2.339.

In 1622, Oughtred improved Gunter's operation: He laid a second scale below the original one. By moving the second scale, he computed without use of a compass. Thus, Oughtred had invented the *slide rule*.

The next 350 years brought numerous improvements of Oughtred's two inventions.[101] An example is Thacher's slide cylinder of 1890.

Thacher's cylindrical slide rule.[102]

It has one scale on the inner rod and a second one on the outer slats. The scales weave back and forth along the rod and the strips.

The overall length of the device is a hefty 61 cm (24 inches). The reward is a huge total scale length of 9 m (10 yards).

You think this is a long scale? A slide cylinder has been proposed[103] where the scale has a record-breaking length of 2 km (1.25 miles).

A modern example of the circular slide rule is the pocket-watch calculator KL-1. Shown here are front and back.

Pocket-watch calculator KL-1.[104]

The scale on one side is fixed, while the scale on the other side can be rotated by one of the two knobs. The two pointers of the scales are connected and are jointly rotated by the second knob.

The design essentially consists of two Bürgi rings of black numbers that are mounted back-to-back. The two connected pointers permit the transfer of values from one side to the other.

The production of slide rules, circular slide rules, and slide cylinders came to a halt in 1976, when the first low-cost electronic pocket calculator entered the market.

———————

It is fun to consider how Bürgi could have designed the first circular slide rule with small computational effort.

12

Design of a Circular Slide Rule

Design and implementation of a circular slide rule would have been easy for Bürgi. We sketch the steps.

Spacing of Ticks

The spacing of ticks on a slide rule cannot go below a minimum that is mandated by the production technology and the capacity of the human eye. The ticks are then so chosen that the minimum spacing is approximately attained.

For any logarithmic scale, this necessarily means that the small numbers of the range 1.0–10.0 have a more detailed representation than the large ones.

Once these numbers have been decided, one looks up the corresponding red numbers in Bürgi's table of logarithms. Below, we use the scaled table for this.

The selection doesn't have to be done with high precision: One simply finds the red number whose black number is closest to the given number.

Let's look at an example.

If we select for the ticks of the slide rule the numbers 1.00, 1.02, 1.04, 1.06, ..., 9.90, 9.95, 10.0, then the corresponding red numbers taken from Bürgi's scaled table are **0, 198, 392, 583, ... 22 927, 22 977, 23 027**.

We get the tick positions by dividing each red number by 23 027 and multiplying with the circumference c of the computing disk. We use Bürgi's scaled table to carry out the computations: We obtain the logarithm of each red number, add the logarithm of $c/23027$, and look up the black number for that sum.

Construction Effort

Assuming an average gap of roughly 60, less than 400 red values have to be processed. This would provide the precision of an engineering slide rule of the 20th century.

While computation of the position of the ticks via Bürgi's table of logarithms would have been easy, the subsequent construction of the circular slide disk would have been a demanding task.

If done with brass sheet-metal, it requires careful etching and labeling. But given the precision with which craftsmen worked in Bürgi's time, it probably could have been done in a few days.

So if Bürgi had cared to pursue this idea, he would have had the first circular slide rule in hand within a week's time.

If he had demonstrated it to scientists and engineers, they would have clamored to obtain such a tool for their work.

During those early uses of the slide disk, Bürgi surely would have come up with modifications for enhanced computations.

Improvements

He would have immediately seen that computation of powers and extraction of roots could be accommodated by addition of a linear scale. Initially he might even have thought to add a linear scale that covers the range **1–23027** of the red numbers of his table.

But then he would have modified that idea: Why not use the linear scale from 1.0 to 10.0?

Indeed, it not only would have been easier to implement, but—more importantly—would have supported simpler computation of powers and extraction of roots than afforded by his table.[105]

The idea of a linear scale from 1.0 to 10.0 is of course the geometric representation of the logarithm with base 10. And that notion would have prompted him to think about constructing a precise table of that logarithm.

He wouldn't have stopped there. Why not add another scale displaying the very precise values he had computed for the sine function?

And then there are numerous applications calling for special scales. They essentially represent particular computations, such as multiplication by $\pi = 3.1415\ldots$ Given Bürgi's background as a craftsman, he quickly would have grasped the potential for such variations of the circular slide disk.

Amazing that he missed all this, just because he didn't exploit the idea inherent in the ring of black numbers of the title page.

So why didn't he pursue this?

Bürgi's Decision

We can only guess at Bürgi's reasoning. But the following seems likely.

A major scientific task at the time was determination of the movements of the stars and planets. It required utter precision of measurements and computations if the results were to decide which of the proposed theories was correct.

Bürgi was a perfect fit for this program: He had the drive and determination to build efficient and precise computational methods, clocks, instruments, and tools.

A circular slide rule would have been inherently imprecise for any reasonable diameter of the disk, and thus wouldn't have lived up to Bürgi's standard. It is likely that this thought convinced Bürgi not to pursue construction of that device.

We move on from Bürgi to Napier, who also constructed a table of logarithms, though with a quite different approach.

13

John Napier

John Napier (1550–1617) began to attend St. Salvator's College in St. Andrews at age 13. He also studied in continental Europe, returned to Scotland in 1571 at age 21, and bought a castle in Gartness.

Upon the death of his father, Sir Archibald Napier, in 1608, Napier moved back to the family's Merchiston Castle in Edinburgh.[106]

He was a mathematician, physicist, and astronomer. Of particular interest was simplification of tedious manual arithmetic.

Toward that goal, he invented several tools.

One of them became known as *Napier's Bones*. They were rods with numbers that, laid side by side, turned multiplication by a single digit number into an almost trivial task.[109]

John Napier.[107]

Napier's Bones, 18th century.[108]

Repeated application of the process plus addition of the results handled multiplication of general numbers.

Napier's tools spawned the development of more complex computing devices that in turn led to mechanical calculators.

The most important contribution is his table of logarithms, which is based on a unique geometric model. Due to this feature, the work is clearly independent from that of Bürgi.

Napier's Model

Consider a line segment L of length N.

A point starts travel at one end of L with initial speed N. Immediately after the departure the speed is reduced. Specifically, when the point has traveled distance y, the speed has been reduced to $N - y$.

We want to compute the time it takes for the point to travel an arbitrary distance $z < N$.

$$0 \text{------} y \text{------} z \text{------} N$$

Denote that travel time by $T(z)$. Using calculus and the natural logarithm ln with base $e = 2.71828\ldots$, we have[110]

$$T(z) = \ln(N) - \ln(N - z)$$

Consider a second point that starts movement from the endpoint of a separate ray at the same time as the first point starts on L, but travels with constant speed N.

At time $T(z)$, the second point has traveled a certain distance $D(z)$.

$$0 \text{------} D(z) \text{ at time } T(z)$$

Since the speed of the second point is N,

$$D(z) = N \cdot T(z) = N(\ln(N) - \ln(N - z))$$

Napier defines $D(z)$ to be the *logarithm* of $N - z$, denoted below by $\text{LOG}_{\text{Napier}}$.

Let $x = N - z$, the distance the first point has yet to cover on the line segment L when it has already traveled distance z. Substituting x for $N - z$ in the above formula for $D(z)$, we have

$$\text{LOG}_{\text{Napier}}(x) = N(\ln(N) - \ln(x)) = N \cdot \ln(N/x)$$

The above relationship between $\text{LOG}_{\text{Napier}}$ and ln was established much later when calculus had been invented and the number e and the related logarithm had been defined.

Indeed, Gottfried Wilhelm Leibniz (1646–1716) and Isaac Newton (1643–1727) independently created calculus more than fifty years later, during the period 1666–1684, and Jacob Bernoulli (1655–1705) defined the number e in 1685.

This implies, of course, that Napier did not compute his table of logarithms using the above formulas. Instead, he used a discrete approximation with $N = 10^7$.

In a multi-year effort, Napier assembled a table of his logarithms in a book titled *Mirifici Logarithmorum Canonis Descriptio*, which one may translate as "Description of the Marvelous Rule of Logarithms."

The table is specifically designed for arithmetic involving the sine values of trigonometry. Toward that goal, the book contains detailed explanations of the underlying mathematics and is filled with example cases to help the user.

Napier's *Mirifici Logarithmorum Canonis Descriptio*, 1614.[111]

The book was printed in 1614, six years prior to the publication of Bürgi's table. It earned Napier fame as the sole inventor of the concept of logarithm. That evaluation persisted until the 19th century.

For the reader interested in details, there is an extensively annotated translation of the book into English.[112] We rely on it next.

The preface of the book outlines the amazing simplification of mathematical operations afforded by the novel concept of logarithm.

The translation of the preface is too long to be listed here. But we include the portion covering the following highlighted segment.[113]

IN MIRIFICVM
Logarithmorum Canonem
Præfatio.

Vum nihil fit/chariſſimi mathematum cultores) mathematice praxi tam moleſtum, quodq; Lo. giſtas magis remoretur, ac retardet, quàm mag. norum numerorū multiplicationes, partitiones, quadratæque ac cubicæ extractiones, quæ præ. ter prolixitatis tædium, lubricis etiam erroribus plurimum ſunt obnoxiæ: Cœpi igitur animo revolvere, quâ arte certâ & expeditâ, poſſem dictis impedimenta amoliri. Multis ſolſ. in de in tunc ſinceo perpenſis, non nulla tandem inveni præcla. ra compendia, alibi fortaſſe tractanda: verùm inter omnia nullum hoc utilius, quod una cum multiplicationibus, par. titionibus, & radicum extractionibus arduis & prolixis ipſos etiam numeros multiplicandos, dividendos, & in radices re. ſolvendos ab opere rejicit, & eorum loco alios ſubſtituit nu. meros, qui illorum munere funguntur per ſolas additiones, ſubſtractiones, bipartitiones, & tripartitiones. Quod quidem arcanum, cùm (ut cætera bona) ſit, quo communius, eo me. lius: in publicum mathematicorum uſum propalare libuit. Eo itaque liberè fruamini (matheſeos ſtudioſi) & quâ à me profectum eſt benevolentiâ, accipite. Valete.

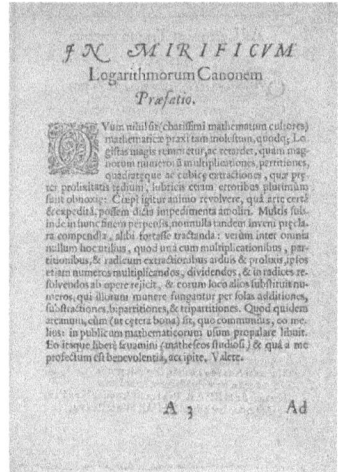

A 3   Ad

etiam numeros multiplicandos, dividendos, & in radices re. ſolvendos ab opere rejicit, & eorum loco alios ſubſtituit nu. meros, qui illorum munere fungantur per ſolas additiones, ſubſtractiones, bipartitiones, & tripartitiones. Quod quidem arcanum, cùm (ut cætera bona) ſit, quo communius, eo me. lius: in publicum mathematicorum uſum propalare libuit. Eo itaque liberè fruamini (matheſeos ſtudioſi) & quâ à me profectum eſt benevolentiâ, accipite. Valete.

Preface of *Mirifici Logarithmorum Canonis Descriptio*, 1614.[114]

"Since indeed the secret [of Napier's invention of the logarithm] is best made common to all, as all good things are, then it is a pleasant task to set out the method for the public use of mathematicians. Thus, students of mathematics, accept and freely enjoy this work that has been produced by my benevolence. Farewell."

These sentences demonstrate the generosity of Napier as he shares his invention without thought of remuneration, and are in stark contrast with the way Bürgi handled publication of his table of logarithms.

Suppose we divide the numbers x as well as the logarithm values of Napier's table by N, getting

$$y = x/N$$
$$\log_{\text{Napier}}(y) = \text{LOG}_{\text{Napier}}(x)/N$$

The earlier equation $\text{LOG}_{\text{Napier}}(x) = N \cdot \ln(N/x)$ then implies

$$\log_{\text{Napier}}(y) = \ln(1/y)$$
$$y = (1/e)^{\log_{\text{Napier}}(y)}$$

Thus, y and $\log_{\text{Napier}}(y)$ constitute a table of logarithms of base $1/e = 0.36787\ldots$

Napier didn't know about this link of his logarithm to $1/e$ and the natural logarithm since neither had been defined at his time. We display this well-known connection only to prove that Napier's logarithm behaves exactly as claimed. That is, addition of logarithm values handles multiplication, subtraction handles division, and so on.

Except for this demonstration, we won't make use of the link to $1/e$ and the natural logarithm. Thus, we follow the spirit of our investigation: We use modern tools to understand relationships, but interpret Napier's actions only using concepts available at that time.

The data of Napier's original table are arranged in an intriguing fashion.

Format of Napier's Table

Napier wanted his table to be particularly useful for arithmetic involving the values of the sine function of trigonometry. Accordingly, the x values are equal to $\sin(\alpha)$ where the angle α ranges from 0 to 90 degrees in 1 minute increments.

Napier arranged the listing of angles, sine values, and logarithms on the pages of the table in a peculiar way.

On any given page, three columns on the left-hand side cover angles α with the corresponding values $x = \sin(\alpha)$ and logarithms $\mathrm{LOG}_{\mathrm{Napier}}(x)$, while three columns on the right hand have the data for angles $90 - \alpha$. As a consequence, the left-hand columns of the pages collectively cover degrees 0 to 45, while the right-hand columns handle degrees 45 to 90 in reverse order.

On the example page of the table displayed next, the angles α on the left start at the top at 5 deg and 30 min and finish at the bottom of the page—not shown—with 5 deg and 60 min. The decreasing angles $90 - \alpha$ on the right begin at the top with 84 deg and 30 min and end at the bottom with 84 deg and 0 min.

Gr.	5						
5 min.	Sinus	Logarithmi	Differentie	logarithmi	Sinus		
30	958458	23450143	23403999	46144	9953962	30	
31	961354	23419980	23373556	46424	9953683	29	
32	964249	23389908	23343203	46705	9953403	28	
33	967144	23359927	23312940	46987	9953122	27	
34	970039	23330036	23282766	47270	9952840	26	
35	972934	23300235	23252681	47554	9952557	25	
36	975829	23270525	23222686	47839	9952274	24	
37	978724	23240903	23192778	48125	9951990	23	
38	981619	23211368	23162956	48412	9951705	22	
39	984514	23181920	23133220	48700	9951419	21	
40	987408	23152560	23103572	48988	9951132	20	
41	990303	23123287	23074010	49277	9950844	19	

Partial page of *Mirifici Logarithmorum Canonis Descriptio*.[115]

For example, for angle α equal to 5 deg and 32 min the table shows $\sin(\alpha) = x = 964\,249$, and $\mathrm{LOG}_{\mathrm{Napier}}(x) = 23\,389\,908$.

Why did Napier select such a strange ordering of the angles, and what is the role of the highlighted Differentia column in the center of the table?

Role of Differentia Column

The Differentia column displays the difference of the logarithm value on the left minus the logarithm value on the right.

That column and the strange ordering of table values resulted from a brilliant idea: Napier wanted to provide logarithms not just for sine values, but also for tangent values.

Since

$$\tan \alpha = \sin(\alpha) / \sin(90 - \alpha)$$

we have

$$\mathrm{LOG}_{\mathrm{Napier}}(\tan(\alpha)) = \mathrm{LOG}_{\mathrm{Napier}}(\sin(\alpha)) - \mathrm{LOG}_{\mathrm{Napier}}(\sin(90 - \alpha))$$

The right-hand side of the equation is the Differentia value, which thus supplies the logarithm of the tangents of angles.

We define a scaled version from Napier's table that we use in the next chapter.

Napier's Scaled Table

Earlier we obtained y and $\log_{\text{Napier}}(y)$ from x and $\text{LOG}_{\text{Napier}}(x)$ by dividing the latter two terms by $N = 10^7$. We apply this scaling to the corresponding entries of Napier's table and thus obtain *Napier's scaled table*.

Put differently, we obtain the scaled table from the original one by moving the decimal point of all x and $\text{LOG}_{\text{Napier}}(x)$ entries 7 positions to the left.

For one thing, this scales the sine values listed in Napier's table as integers x ranging from 0 to 10 000 000 to the modern values y ranging from 0.0 to 1.0.

We briefly discuss the accuracy of the original table and thus of the scaled version.

The earlier-cited annotated translation[116] of *Mirifici Logarithmorum Canonis Descriptio* points out numerous instances where the published value differs to some degree from the correct value of the logarithm of base $1/e$.

We ignore this aspect and always use the original values. This simplifies the presentation and avoids possible confusion whether cited values occur in the original table or are corrected versions.

———————

In the next chapter we see how Napier's scaled table can be used for computation outside trigonometry.

14

Computation with Napier's Table

Suppose we wanted to use Napier's table of logarithms for general computations and not just those involving sine values.

This would require some care for the following reasons.

First, the sine values of very small angles are separated by relatively large differences, as demonstrated by the left portion of the first page of the table displayed here.

Second, the $\text{LOG}_{\text{Napier}}(x)$ function is very nonlinear in that region.

The combined effect is that interpolation in that region would produce quite imprecise estimates.

Napier was well aware of this and in his instructions recommended a certain shifting of logarithm values.

Gr.		0	
'o			
min	*Sinus.*		*Logarithmi*
☀	o		infinitum
1	2909		81425681
2	5818		74494213
3	8727		70439564
4	11636		67562745
5	14544		65331315
6	17453		63508099
7	20362		61966595
8	23271		60631284
9	26180		59453453
10	29088		58399857
11	31997		57446759
12	34906		56576646
13	37815		55776222
14	40724		55035148
15	43632		54345225
16	46541		53699843
17	49450		53093600
18	52359		52522019
19	55268		51981356
20	58177		51468431
21	61086		50980537
22	63995		50515342
23	66904		50070827

Mirifici Logarithmorum Canonis Descriptio. Initial part of table.[117]

Such shifting is handled implicitly when one uses the scaled version of Napier's table defined in Chapter 13.

Recall that the scaled table is obtained from the original one by shifting the decimal point of the x and $\text{LOG}_{\text{Napier}}(x)$ entries 7 positions to the left.

We modify the scaled table a bit to simplify its use.

First, we insert into the table a new entry with $y = 0.1$ lying between $y = 0.099\,8987$ of angle $\alpha = 5$ deg 44 min and $y = 0.100\,1881$ of angle $\alpha = 5$ deg 45 min. Napier's instructions supply for $y = 0.1$ the logarithm value[118] **2.302 5842**. We use the bold font for logarithms in the spirit of the red numbers of Bürgi's table. Indeed, for consistency we take the bold font to represent red color here even though Napier never declared his logarithm values to have that color.

Second, we discard all entries preceding the added entry, and thus end up with a table having logarithm values for y values ranging from 0.1 to 1.0.

Below, we still refer to this modified version as *Napier's scaled table* without risk of confusion. It is the analog of Bürgi's scaled table.

Let's see how we can use the scaled table to simplify computations.

Using Napier's Scaled Table

The y values of the scaled table have reasonably small spacing throughout, and the corresponding logarithm values range from **2.302 5842** to **0**.

Here is the beginning and end of the scaled table, plus a few intermediate entries that demonstrate the claimed uniformly small spacing.

y	$\log_{\text{Napier}}(y)$
0.1 000 000	**2.302 5842**
0.100 188 1	**2.300 7056**
0.100 477 5	**2.297 8212**

. . .

$$
\begin{array}{ll}
0.510\,5430 & \textbf{0.672\,2802} \\
0.510\,7932 & \textbf{0.671\,7905} \\
0.511\,0431 & \textbf{0.671\,3011}
\end{array}
$$

$$\cdots$$

$$
\begin{array}{ll}
0.999\,9996 & \textbf{0.000\,0004} \\
0.999\,9998 & \textbf{0.000\,0002} \\
1.000\,0000 & \textbf{0.000\,0000}
\end{array}
$$

Computation

Multiplication, division, computing of powers, and extraction of roots are done analogously to the Bürgi case.

In the discussion of these operations we denote the base underlying the logarithm values by b. From the preceding chapter, we know that $b = 1/e = 0.36787\ldots$

As pointed out in Chapter 13, Napier didn't know about the connection of his logarithm with $1/e$ since e hadn't been defined yet. For consistency, we don't utilize the numerical value of b below either. Thus, he could have carried out computations with the scaled table as described below, just as Bürgi could have done so with Bürgi's scaled table.

Initial scaling with powers of 10 moves a given number into the range 0.1–1.0.

The entries 0.1 and **2.302 5842** in the first row of the table effectively say that

$$b^{2.302\,5842} = 0.1$$

Let's call the logarithm value **2.302 5842** the *Napier constant*.

The Napier constant and scaling with powers of 10 are linked as follows.

For any k, the value of any b^n doesn't change if we multiply it with 10^k and add $k \cdot \textbf{2.302\,5842}$ to the exponent n. Thus we have for any integer k the *Napier scaling*

$$b^n = 10^k \cdot b^{n+k \cdot 2.302\,5842}$$

We illustrate these operations with a summary of the multiplication case.

Multiplication

We want to compute $133.1 \cdot 4.022$. Throughout, we round logarithms to four significant digits for clarity.

$$
\begin{aligned}
133.1 \cdot 4.022 &= 10^4 \cdot 0.1331 \cdot 0.4022 && \text{(initial scaling)} \\
&= 10^4 \cdot b^{2.016} \cdot b^{0.911} && \text{(scaled table)} \\
&= 10^4 \cdot (b^{2.016 + 0.911}) && \text{(addition)} \\
&= 10^4 \cdot b^{2.927} && \text{(simplify)} \\
&= 10^4 \cdot 10^{-1} \cdot b^{2.927 - 2.303} && \text{(N. scaling } k = -1) \\
&= 10^3 \cdot b^{0.624} && \text{(simplify)} \\
&= 10^3 \cdot 0.5358 = 535.8 && \text{(scaled table)}
\end{aligned}
$$

The final results 535.8 is a bit different from the correct 535.3282 due to rounding of logarithm values and minor errors in Napier's original table, which thus are also present in the scaled table.

Division, Powers, Roots

Division, computation of powers, and extraction of roots can be derived in similar fashion from the Bürgi examples in straightforward fashion. Here, too, the initial scaling has the beneficial aspect that Napier scaling for multiplication and division isn't needed at all or involves 10^k with $k = \pm 1$.

In 1616, Henry Briggs (1561–1630) visited Napier and proposed construction of a new table of logarithms that would be easier to use. Napier immediately agreed—another sign of his generous nature—and Briggs started the computations that year.

15

Henry Briggs

Henry Briggs (1561–1630) started his education at St. John's College in Cambridge during the years 1577–1585. He advanced to the position of examiner and lecturer in the faculty in 1592.

In 1596 he became the first professor of geometry at the recently founded Gresham College in London. Besides mathematics, he taught astronomy and navigation. As we saw in Chapter 11, he was first a tutor and later a colleague of Gunter.[119]

During his visit of Napier in 1616, Briggs proposed 10 as base for a new table of logarithms. We denote that logarithm by $\log_{\text{Briggs}}$. By 1617, Briggs had computed the logarithms for the numbers 1–1000 and published them in *Logarithmorum Chilias Prima*.[120]

Seven years later, in 1624, he published a much larger table in *Arithmetica Logarithmica*.[122]

On the title page Briggs takes great care to state that Napier was the sole inventor of logarithms. That conclusion would stand until the 19th century, when Bürgi's work became more widely known.

Briggs's *Logarithmorum Chilias Prima*, 1617.[121]

numeris naturali ſerie creſcentibus ab vnitate ad
10,000 : et a 90,000 ad 100,000. Quorum ope multa
perficiuntur Arithmetica problemata
et Geometrica.

HOS NVMEROS PRIMVS
INVENIT CLARISSIMVS VIR IOHANNES
NEPERVS Baro Merchiſtonij : eos autem ex eiuſdem ſententia
mutavit, eorumque ortum et vſum illuſtravit HENRICVS BRIGGIVS,
in celeberrima Academia Oxonienſi Geometriæ
profeſſor SAVILIANVS.

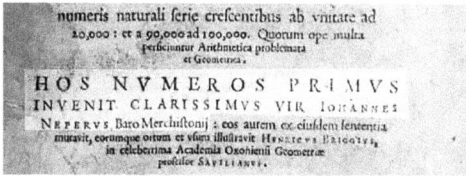

Briggs's *Arithmetica Logarithmica*, detail.

The highlighted part states

"Which numbers [the logarithms] invented the brilliant gentleman John Napier, Baron of Merchiston."

The book describes on 88 pages the computational methods, and on 289 pages 14-digit $\log_{\text{Briggs}}$ values for the numbers 1–20 000 and 90 000–100 000, for a total of 30 000 entries.

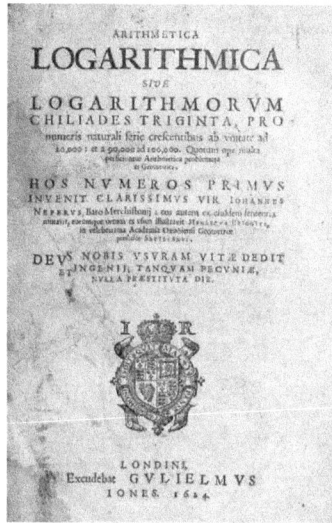

ARITHMETICA
LOGARITHMICA
SIVE
LOGARITHMORVM
CHILIADES TRIGINTA, PRO
numeris naturali ſerie creſcentibus ab vnitate ad
10,000 : et a 90,000 ad 100,000. Quorum ope multa
perficiuntur Arithmetica problemata
et Geometrica.

HOS NVMEROS PRIMVS
INVENIT CLARISSIMVS VIR IOHANNES
NEPERVS, Baro Merchiſtonij : eos autem ex eiuſdem ſententia
mutavit, eorumque ortum et vſum illuſtravit HENRICVS BRIGGIVS,
in celeberrima Academia Oxonienſi Geometriæ
profeſſor SAVILIANVS.

DEVS NOBIS VSVRAM VITÆ DEDIT
ET INGENIJ, TANQVAM PECVNIÆ,
NVLLA PRÆSTITVTA DIE.

I R

LONDINI
Excudebat GVLIELMVS
IONES 1624

Briggs's *Arithmetica Logarithmica*, 1624.[123]

Chilias tertia.

Num. abſolu	Logarithmi.	Num. abſolu	Logarithmi.	Num. abſolu	Logarithmi.
2701	3,43152,45841,8745 16,07604,9856	2734	3,43679,85102,3181 15,88104,3764	2767	3,44200,91591,4095 15,69266,4377
2702	3,43168,53446,8601 16,07010,1271	2735	3,43695,73306,6945 15,87613,7863	2768	3,44216,60857,8472 15,68699,6085
2703	3,43184,60456,9872 16,06415,7088	2736	3,43711,60930,4808 15,87043,6205	2769	3,44232,29557,4557 15,68133,1887
2704	3,43200,65872,6960 16,05821,7299	2737	3,43727,47974,1013 15,86463,8785	2770	3,44247,97690,6444 15,67567,1779
2705	3,43216,71694,4259 16,05228,1901	2738	3,43743,34437,9798 15,85884,5599	2771	3,44263,65157,8223 15,67001,5754
2706	3,43232,77922,6160 16,04635,0891	2739	3,43759,20322,5397 15,85305,6641	2772	3,44279,32159,3977 15,66436,3809
2707	3,43248,82557,7051 16,04042,4260	2740	3,43775,05628,2039 15,84727,1911	2773	3,44294,98695,7786 15,65871,5941
2708	3,43264,86600,1311 16,03450,2005	2741	3,43790,90355,3950 15,84149,1400	2774	3,44310,64567,3727 15,65307,2143
2709	3,43280,90050,3316 16,02858,4124	2742	3,43806,74504,5350 15,83571,5104	2775	3,44326,29874,5870 15,64743,2412
2710	3,43296,92908,7440 16,02267,0608	2743	3,43822,58076,0454 15,82994,3018	2776	3,44341,94617,8282 15,64179,6744
2711	3,43312,95175,8048 16,01676,1454	2744	3,43838,41070,3472 15,82417,5139	2777	3,44357,58797,5026 15,63616,5514
2712	3,43328,96851,9502 16,01085,6657	2745	3,43854,23487,8611 15,81841,1461	2778	3,44373,22414,0160 15,63053,7177
2713	3,43344,97937,6159 16,00495,6211	2746	3,43870,05329,0073 15,81265,1982	2779	3,44388,85467,7737 15,62491,4071
2714	3,43360,98433,2371 15,99906,0115	2747	3,43885,86594,2055 15,80689,6696	2780	3,44404,47959,1808 15,61929,4668
2715	3,43376,98339,2486 15,99316,8360	2748	3,43901,67283,8751 15,80114,5596	2781	3,44420,09888,6416 15,61367,9187
2716	3,43392,97656,0846 15,98728,0943	2749	3,43917,47398,4347 15,79539,8679	2782	3,44435,71256,5603 15,60806,7802
2717	3,43408,96384,1789 15,98139,7858	2750	3,43933,26938,3026 15,78965,5941	2783	3,44451,32063,3405 15,60246,0448
2718	3,43424,94523,9647 15,97551,9103	2751	3,43949,05903,8968 15,78391,7379	2784	3,44466,92309,3853 15,59685,7122
2719	3,43440,92075,8750	2752	3,43964,84295,6347	2785	3,44482,51995,0975

Partial page of logarithm table of *Arithmetica Logarithmica*.[124]

Since the table doesn't contain $\log_{\text{Briggs}}$ values for the numbers $20\,001 - 89\,999$, isn't it incomplete and thus not generally usable?

A short answer is an emphatic "No." A longer answer is included below, once we have restructured the table a bit.[125]

Let's look at the $\log_{\text{Briggs}}$ values for 2, 20, and 200. They are

$$0.301\,029\,995\,663\,98$$
$$1.301\,029\,995\,663\,98$$
$$2.301\,029\,995\,663\,98$$

They differ only in the whole number. Indeed, if we scale up any decimal number y by a factor 10^k, then $\log_{\text{Briggs}}(y)$ changes to

$$\log_{\text{Briggs}}(10^k \cdot y) = \log_{\text{Briggs}}(y) + \log_{\text{Briggs}}(10^k)$$
$$= \log_{\text{Briggs}}(y) + k$$

This is Briggs's ingenious use of the way the decimal numbers are connected with their base 10 logarithms. One could call this simple addition step *Briggs scaling*. It corresponds to Bürgi and Napier scaling.

We use that fact to restructure Briggs's table so that it doesn't contain redundant listings of numbers that differ only by powers of 10 and whose $\log_{\text{Briggs}}$ values therefore differ only in the whole number preceding the decimal point.

First, we delete all numbers whose lowest digit is 0, together with their $\log_{\text{Briggs}}$ values. This is appropriate since the table also contains a number without that 0, which thus differs only by a power of 10.

As an aside: How many number/logarithm pairs have we deleted? A simple calculation concludes that about 4 000 numbers have been removed. Hence the reduced table contains about 26 000 entries.

Why did Briggs introduce such redundancy?

Since Descartes's notation for exponents didn't exist yet, one may conjecture that Briggs felt compelled to provide a table that could be used without scaling.

Second, we insert into each of the remaining numbers a decimal point after the leftmost digit and replace the leftmost digit of each $\log_{\text{Briggs}}$ value by 0.

For example, 205 becomes 2.05, and the associated $\log_{\text{Briggs}}$ value 2.31175... becomes 0.31175...

Third, we sort the entire list of number/logarithm pairs in ascending order. The resulting numbers go from 1.0 to 9.9999 and the associated logarithm values from 0.0 to 0.999 995 657 033 47. Let's call the resulting list of numbers and their $\log_{\text{Briggs}}$ values *Briggs's scaled table*.

Briggs's Scaled Table

Remember that Briggs's table doesn't have logarithms for the numbers 20 001–89 999?

This doesn't mean that Briggs's table cannot be used for general computations. It simply implies that the numbers beyond 2.0 and below 9.0 in the scaled table are based on entries of the original table in the range 2 001–8 999.

In the scaled table this manifests itself as follows. Successive numbers in the range 2.0 to 9.0 differ by 0.001, while the remaining successive numbers differ by the smaller 0.0001.

Accordingly, the precision of interpolation in the scaled table varies and is highest at the beginning and end of the table.

Here is an excerpt. We use bold font to simulate red color to be consistent with the display of Bürgi's and Napier's scaled tables.

y	$\log_{\text{Briggs}}(y)$
1.0000	**0.000 000 000 000 00**
1.0001	**0.000 043 427 276 87**
	...
2.0000	**0.301 029 995 663 98**
2.001	**0.301 247 088 636 21**

$$\cdots$$

5.000	0.698 970 004 336 02
5.001	0.699 056 854 547 66

$$\cdots$$

8.999	0.954 194 251 815 87
9.000 0	0.954 242 509 439 32

$$\cdots$$

9.999 9	0.999 995 657 033 47
10.000 0	1.000 000 000 000 00

Briggs was well aware of that flaw. But after years of computing effort, he felt unable to complete the computations filling that gap. He prepared detailed instructions for how to calculate these numbers, but nobody volunteered to take on the significant computational burden.[126]

Computation

Multiplication, division, and computation of powers with the scaled table are so simple that we skip examples. We only mention that *initial scaling* moves the given numbers inside the range 1.0–10.0.

We include details for the extraction of roots. Say the k-th root of some number is to be taken.

Do an initial scaling with powers of 10^k such that a number strictly between 1 and 10^k is achieved. Suppose the scaling factor is $10^{k \cdot m}$. Then the scaled number, say x, is, for some $r < k$ and $1.0 \leq y < 10.0$, equal to $10^r \cdot y$.

Using the scaled table, obtain $\log_{\text{Briggs}}(y)$. Thus, $\log_{\text{Briggs}}(x) = \log_{\text{Briggs}}(y) + r$.

The kth root of x, say z, has $\log_{\text{Briggs}}(z) = \log_{\text{Briggs}}(x)/k$. Since $r < k$, that logarithm value is a proper fraction, and z can be directly looked up in the scaled table.

Then $10^m \cdot z$ is the kth root of the original number.

We solve an example problem that we treated earlier with Bürgi's table. We use the same notation, but replace Bürgi's base 1.0001 by 10. We also round the logarithm values of Briggs's scaled table.

$$
\begin{aligned}
\sqrt[6]{4.05006 \cdot 10^{14}} &= 10^2 \cdot \sqrt[6]{4.05006 \cdot 10^2} && \text{(initial scaling)} \\
&= 10^2 \cdot \sqrt[6]{10^{0.60746} \cdot 10^2} && \text{(scaled table)} \\
&= 10^2 \cdot \sqrt[6]{10^{0.60746+2}} && \text{(add)} \\
&= 10^2 \cdot \sqrt[6]{10^{2.60746}} && \text{(simplify)} \\
&= 10^2 \cdot 10^{2.60746/6} && \text{(division by 6)} \\
&= 10^2 \cdot 10^{0.43458} && \text{(simplify)} \\
&= 2.720 \cdot 10^2 = 272.0 && \text{(scaled table)}
\end{aligned}
$$

At this point we have insight into the three types of logarithms created by Bürgi, Napier, and Briggs. Indeed, the three scaled tables derived from their results exhibit the common thread by which these logarithms, which look so different on the surface, are connected.

In the next chapter we compare the accuracy and computational efficiency afforded by these scaled tables.

16

Comparison of Accuracy and Efficiency

We estimate the accuracy and efficiency afforded by the scaled tables of Bürgi, Napier, and Briggs.[127]

Accuracy

Just as we did in Chapter 6, we represent the accuracy approximately by the *inherent accuracy* of each table, which is measured via maximum interpolation errors.

Bürgi's Scaled Table

Chapter 6 establishes the inherent accuracy of Bürgi's scaled table to be 9 digits. The result is based on the maximum interpolation error when a black value is derived from a red logarithm value.

The case of Napier's scaled table is a bit more complicated, as we see next.

Napier's Scaled Table

While Bürgi computes black values for the possible red logarithm values, Napier starts with angles between 0 and 90 degrees in 1 minute increments, computes the corresponding sine values, and then, for the latter numbers, establishes the corresponding $\log_{\text{Napier}}$ values.

Here we view the sine values as the given black numbers and derive the $\log_{\text{Napier}}$ values as the red ones. We then define the *inherent accuracy* of Napier's scaled table to be the precision with which interpolation can accomplish that step.

Napier's original table is based on even spacing of the angles between 0 and 90 degrees in 1 minute increments. As the angles reach 90 degrees, the sine values converge to 1.0 in ever smaller steps.

Correspondingly, the black numbers of the scaled table, which are those sine values, are tightly clustered near 1.0 and therefore have a correspondingly larger spacing near 0.1.

Since the interpolation error near 0.1 is maximum, the increased spacing of black numbers near that point inflates that error.

Hence, we compute the error with and without that inflation effect. The result is that Napier's scaled table has an inherent accuracy of 6 digits that goes up to 7 digits when the inflation effect is removed.

The evaluation of Briggs's scaled table also entails a complication.

Briggs's Scaled Table

We define the *inherent accuracy* of Briggs's table analogous to that for Napier. Thus, the interpolation error when a red logarithm value is computed for a given black one determines the precision.

A complication arises since Briggs's scaled table doesn't have uniform spacing. Indeed, his goal—never attained—was a larger table that upon scaling would have had uniform 0.0001 spacing.

For the scaled table derived from Briggs's table, the inherent accuracy is 8 digits and for the larger scaled table 9 digits.

Size of Table Entries

It is very likely that Bürgi predicted the computational perfor-
mance and thus chose 9 digit representation for the table entries.
The same applies to Napier's table, since the logarithms are listed
with 7 digits, and we have computed 6 or 7 digit accuracy.

On the other hand, it is odd that Briggs chose 14 digits precision
for the table, yet inherent accuracy is only 9 digits even if the table
had been extended to a uniform spacing of 0.0001. Indeed, it would
seem that, say, 10 digit precision would have done just as well.

Now let's look at the efficiency with which computations can be
carried out.

Comparison of Efficiency

Multiplication, division, computation of powers, and extraction of
roots require similar effort when the scaled tables of Bürgi and
Napier are used. This is evident from the summarizing equations
depicting the computational steps.

Indeed, both scaled tables require manipulation of the Bürgi or
Napier constant to bring logarithm values into the range of the
table. That adjustment is trivial for Briggs's scaled table. In fact,
that difference makes the use of Briggs's scaled table much more
efficient.

The interpolation effort varies. One may be tempted to say that the
interpolation for general computations with Napier's scaled table
is most time-consuming since ratios involving varying differences
of irregularly spaced black and red values must be computed and
manipulated. In addition, the table does not provide those differences.

This harsh criticism is not justified since Napier intended his table
to be mostly used for trigonometric computations. For that task,
the table is ingeniously structured so that sine and tangent values

are simultaneously covered. But it still would have been useful if the table had provided differences of the red logarithm values.

Somewhat easier is the interpolation process for Bürgi's table. Except for a few entries toward the end of the table, successive red logarithms values differ by 1.0, and differences of the black values are implicitly given.

Briggs's table has the same advantage. Here, differences between red logarithm values are explicitly provided, and successive black numbers always differ by 0.001 or 0.0001.

In subsequent centuries, mathematicians computed many additional logarithm tables. We cannot cover those accomplishments here, but acknowledge a Herculean effort of the 19th century.

17

Beyond Bürgi, Napier, and Briggs

In 1848, Edward Sang (1805–1890) of Edinburgh, Scotland and two of his four daughters, Flora Chalmers Sang (1838–1925) and Jane Nicol Sang (1834–1878), started work on a 47-volume masterpiece of logarithm tables.[128] They finished the effort 27 years later. They split the effort: Edward Sang created 26 volumes, Flora Chalmers Sang 16 volumes, and Jane Nicol Sang 5 volumes.

Volumes 1–3 explain the methodology. Volume 4 has 28 digit logarithms for all prime numbers up to 10,000, and a few beyond. Volumes 5 and 6 have 28 digit logarithms for the numbers 1–20,000. Volumes 7–38 contain 15 digit logarithms for the numbers 100 000–370 000. The remaining nine volumes contain tables for trigonometry and astronomy.

The 47 volumes were never printed since production cost would have been huge and demand low.

Edward Sang.[129]

Even before Edward, Flora, and Jane Sang started their effort, a novel invention of mechanized computation was proposed that in principle would create logarithm tables without any human effort.

The Difference Engine

In 1822, the outstanding engineer, inventor, mathematician and philosopher Charles Babbage (1791–1871) designed the *difference engine*.[130]

It reduced the evaluation of polynomials $a_n \cdot x^n + a_{n-1} \cdot x^{n-1} + \ldots + a_1 \cdot x^1 + a_0$ for the values $x = 1, 2, 3, \ldots$ to a series of additions. Specifically the computation of N values required about $n \cdot N$ additions and thus was very efficient. The machine printed tabular output and thus eliminated human, and hence error-prone, transcription of results.

Why did Babbage focus on the evaluation of polynomials?

Charles Babbage.[131]

By his time, it had been established that many functions of interest—such as the logarithm function—could be represented by polynomial functions $a_n \cdot x^n + a_{n-1} \cdot x^{n-1} + \ldots + a_1 \cdot x^1 + a_0$ to any desired degree of accuracy.

Thus, Babbage's difference engine could compute tables for any function represented that way, in particular for any logarithm function.

Babbage could not complete the implementation of the difference engine due to several factors, not the least of which the technical complexity of the machine and the comparatively inefficient tools to manufacture the large number of components.

It is estimated that it would have been composed of around 25 000 parts weighing 15 tons in total.[132]

In the period 1847–49—around the time Edward, Flora, and Jane Sang started their effort—he redesigned the machine and produced *difference engine no. 2*, which was substantially simpler.

It also wasn't constructed during Babbage's lifetime, but a determined effort in the late 20th and early 21st century created two working engines.[133] Each machine has about 8 000 parts and weighs five tons.[134]

The photo shows the engine on view in the Science Museum in London.

Difference engine no. 2. Science Museum, London.[135]

The discussion so far has focused on the construction and use of logarithm tables, and their physical implementation in slide rules, disks, and drums.

The parallel mathematical development[136] eventually led to the modern definition of the logarithm function by Euler[137]: For any basis b, define the logarithm of y by

$$\log_b(y) = x \text{ if } b^x = y$$

We shall not trace that development or its consequences, but mention only that while the logarithm function was first confined to real numbers, beginning with Euler it also involved imaginary and complex numbers.

Among the relationships is Euler's famous equation establishing an astonishing link connecting algebra and geometry using $e = 2.71828\ldots$, $i = \sqrt{-1}$, and the cosine and sine functions $\cos(x)$ and $\sin(x)$ of trigonometry.

$$e^{ix} = \cos(x) + i\sin(x)$$

Applying Euler's definition of the logarithm function and denoting $\log_e$ by ln, we have

$$\ln(\cos(x) + i\sin(x)) = i \cdot x$$

We are coming to the last part of the book, where we deal with the question of who invented logarithms and when.

This is a thorny question since a number of conflicting answers exist. Amazingly, most of them seem well justified.

How is that possible?

In our answer we rely on certain results of brain science and a model-based interpretation of the world.[138]

The next chapter covers this material; the subsequent one uses it to explain how conflicting yet well-justified answers to the priority question are possible.

As you surely expect, we also offer our own answer.

18

Models of the World

Stephan Hawking and Leonhard Mlodinow coined the term *Model-dependent Realism*[139] to capture the notion that scientists use models to understand the world.

They use that concept to justify physics models of the origin of the universe: The theory of the Big Bang and its consequences, depicted here with the timeline and the metric expansion of space.

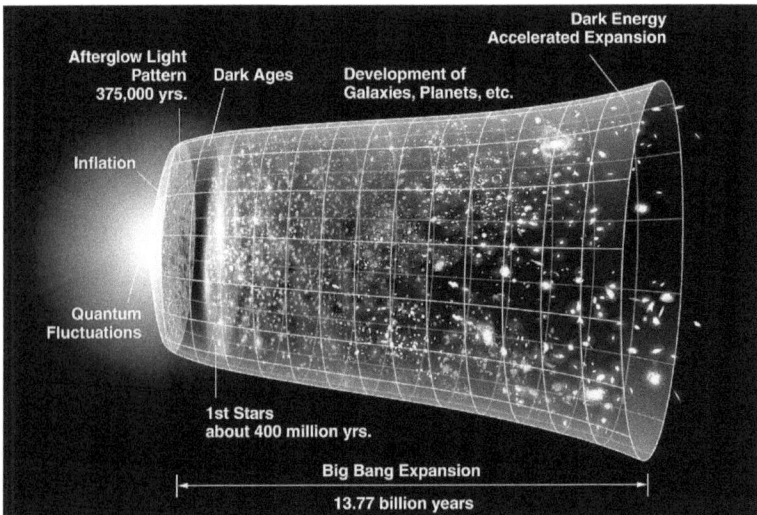

Big Bang model of time and space.[140]

More generally, they say that we, the humans of the earth, not only use models all the time to *understand* the world, but that we postulate them to *be* the world.

As a matter of convenience—though quite misleadingly—we call the conclusions derived from the models *facts* of the world, as if they were conclusions independent of human thought.

Mind you, the main assertion of model-dependent realism—that we model the world and then treat the model results as facts—cannot be proven like a theorem of mathematics.

Instead, it is a *viewpoint* that helps us navigate through and clarify the ideas, proposals, questions, claims, beliefs, superstitions, and hypotheses we encounter every day.

Here is an example of such a clarification.

An Example

A strong believer of a certain religion says, "There are beliefs in life, and there are facts."

We don't know how to cope with "beliefs in life." But we are willing to get into a discussion about the nature of facts.

Hence we ask for an example of a fact. The answer is, "Gravity is a fact. If I lift up a rock and then release it, I know for a fact that it will drop back down. So this is a fact and not a belief."

Is that really the case? All we really know is that, since humanity started to record history a few thousand years ago, there has never been a report where a rock released from a human hand rose upward.

But how do we know that this won't happen some time in the future?

Or as the philosopher Ludwig Wittgenstein put it:[141] "That the sun will rise tomorrow, is a hypothesis; and that means: We do not *know* whether it will rise."

And how can we be sure that the event of an upward rising rock hasn't happened and continues to happen, but with such a rare occurrence—say every 10 000 years—that, even if observed, it would be discounted as a momentary illusion and thus not recorded?

Indeed, quantum physics claims that the event of a rising rock does take place, but in intervals averaging billions of years.

So how do we respond to the claim regarding beliefs and facts?

We walk through the above questions and conclusions, explain the notion of model-dependent realism, and finally show that the claimed fact of gravity actually is one of the basic assumptions of the Newtonian model of the world.

Types of Models

The concept of model isn't just a human creation that handles rocks falling down, water boiling at a certain temperature, the sun rising in the East, or the universe starting with the Big Bang.

Indeed, our brain uses numerous models below the level of consciousness to cope with the world. The results of these models then surface in our consciousness as decisions or facts.

So in some sense evolution has created model-dependent realism within our brains before anybody had that idea. Modern brain science has proved this seemingly fantastic claim of models operated by the brain.[142]

One such result explains why we feel tired.[143]

No, this isn't the body saying that rest is needed. Instead, the brain has computed that efforts at the current level, if continued for an extended period of not just hours but days, would damage the body.

Hence, as a precaution, the brain sends symptoms of fatigue to the level of consciousness so we stop current activities and rest. The symptoms may include yawning, difficulty in keeping our eyes open, the need to sit down, and so on, all with the goal that our conscious decision process becomes convinced that the body should rest.

The reliance on this huge variety of models—from subconscious processes of the brain to abstract ideas about the origin of the universe—also have a downside: Their use invariably results in errors small and large, and sometimes in colossal blunders.

Modeling Errors

For a demonstration of modeling errors at the subconscious level, look at two pictures of a checkered square with white and black tiles, plus a column producing a shadow. Here is the first one.

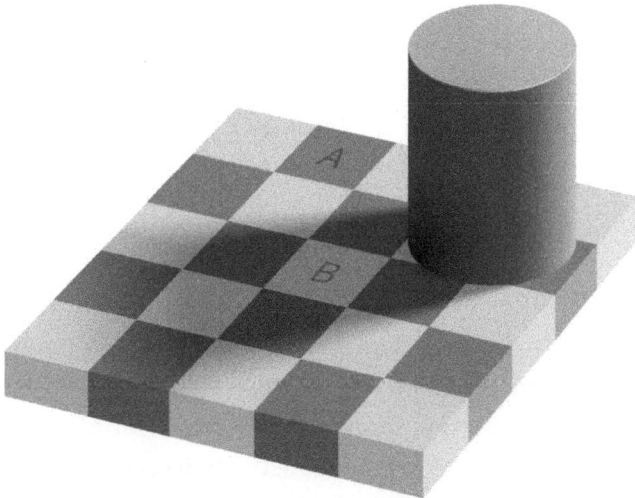

Perception: Square A is much darker than square B.[144]

If you are asked about the color of the two squares labeled "A" in the topmost row and "B" next to the cylinder, you most likely say that square A is black and square B is white.

Or, if you want to be more precise and avoid the extreme terms "black" and "white," you may instead declare that the pixels of A are much darker than those of B.

In the second picture below, a corridor connects the squares A and B that has exactly the same gray shade as both A and B, proving that the pixels of these two squares have the identical color!

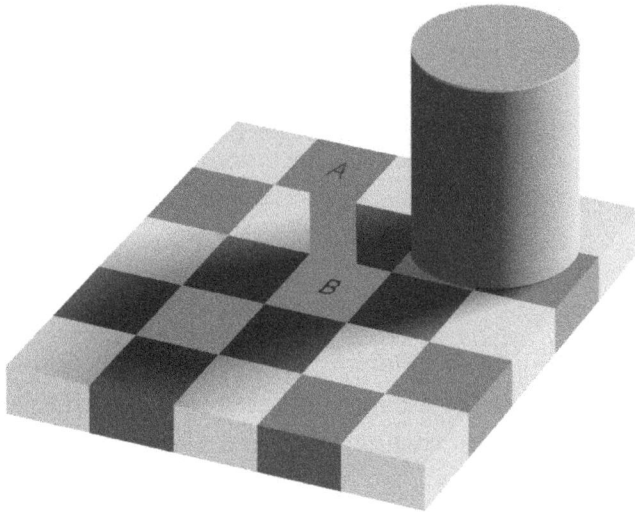

Reality: Squares A and B have the same shade of gray.[145]

No matter how often you go back and forth looking at the two pictures, you cannot shake off the interpretation that the pixels of A in the first picture are much darker than those of B, and that they have the same gray color in the second one.

What is happening here?

In the first picture, the brain uses a model of determining color where the shading effect of the column is taken into account. This occurs at a level below consciousness, so in some sense is hard wired: No matter how you consciously try to counteract this interpretation after seeing the second photo, the brain will insist on that interpretation.

In contrast, in the second picture the brain—again at a level below consciousness—uses a model that essentially says that a contiguous area that looks uniformly colored is indeed uniformly colored. That model is used to evaluate the area composed of A, B, and the connecting corridor.

How can it be that the brain employs such disparate and potentially conflicting models?

Sources of Errors

The explanation is simple: As the brain builds these models starting in infancy or even *in utero*, it cannot possibly check whether they are all consistent. It can only make adjustments when such conflicts become apparent and, more importantly, when it sees a way to adjust at all.

The change can be reasonably accomplished at the level of consciousness, but is much more difficult below that level. The two pictures demonstrate this difficulty. Indeed, after we have gone back and forth between the two pictures a number of times, we simply give up and declare that we cannot train the brain to avoid the conflict.

At the conscious level, errors range from almost irrelevant to significant to catastrophic. Examples are a minor computing error caused by rounding of numbers, an incorrectly constructed calendar leading farmers to plant at the wrong time, and fanatic religious claims that result in the slaughter of millions of people who believe in a different deity.

Before we put the idea of model-dependent realism to use, let's briefly look at true facts.

True Facts

Almost all of modern mathematics consists of true facts. The foundation of that astonishing building of human thought consists of axioms, which are nothing but assumptions. They cannot ever be verified, so they are considered true by declaration.

Since almost all of the building of mathematics is constructed from these axioms via logic—a particular part of mathematics that also rests upon axioms—it almost entirely consists of true facts.

There are more true facts. Examples are statements like, "I have a severe headache" and "To me, the Taj Mahal is the most beautiful building in the world." Indeed, many parts of life that make us human and differentiate us from machines, involve such facts.

So while we focus here on model-dependent realism to understand the question "Who has invented logarithms?", we are well aware that there is much more to life than explaining the world with man-made models.

––––––––––––––––

We are ready to investigate the question "Who invented logarithms?".

19

Who Invented Logarithms?

The question has been answered in several ways.

- Briggs declared 1624 that Napier was the sole inventor.[146] This judgement persisted for centuries despite the next statement.
- Kepler wrote 1626 that Bürgi had assembled the table of logarithms many years before Napier published his table of logarithms.[147]

Since the mid-19th century, there have been additional opinions. Here are three well-reasoned example claims:

- Bürgi and Napier are co-inventors.[148]
- Bürgi constructed a logarithmic table without logarithms, while Napier did use them.[149]
- The logarithmic function really begins with Kepler and culminated in Euler's function definition that allowed extension to complex values.[150]

A knotty situation, wouldn't you agree?

How can we resolve these conflicting assessments?

We invoke the insight of the preceding chapter. In particular, we rely on the notion that the human brain represents many aspects of the world by models, and that the brain declares conclusions drawn via such models to be facts.

We propose that this process creates the confusing situation at hand. The resolution of the conflicting opinions then doesn't come via careful dissection of the arguments producing the varied conclusions. Rather, it comes through a fundamental examination of the underlying models of the world.

By the way, this is a general recipe for clarity when discussions about historical events become heated and do not converge to a mutually agreed-upon interpretation. A major reason for such discord may well be that the various claims and counterclaims implicitly assume different underlying models.

With the modeling aspects in mind, let's try and work toward resolution of the priority question.

A summary of the historical developments sets the stage.

Key Steps

First, Archimedes realizes that multiplication of powers of a given number is achieved by addition of exponents. In Descartes's notation, this means for example that $10^m \cdot 10^n = 10^{m+n}$.

Second, Apollonius of Perga invents exponents for the Greek number M, which stands for $10\,000$.

In terms of equivalent decimal numbers, he creates with the Greek numbers $\alpha = 1, \beta = 2, \gamma = 3, \delta = 4, \ldots$ the numbers

$$\overset{\alpha}{M} = 10\,000^1; \quad \overset{\beta}{M} = 10\,000^2; \quad \overset{\gamma}{M} = 10\,000^3; \quad \overset{\delta}{M} = 10\,000^4; \ldots$$

He surely was aware that addition of the exponents was equivalent to multiplication of the numbers represented with them. For example,

$$\overset{\alpha}{M} \cdot \overset{\gamma}{M} = \overset{\alpha+\gamma}{M} = \overset{\delta}{M}$$

Third, Virasena represents any number x as the product of a power of 2 times an odd number y, then develops rules involving that concept.

Fourth, Stifel displays two sequences of numbers: One sequence has the exponents $\ldots -3, -2, -1, 0, 1, 2, 3 \ldots$, and the other has the powers of 2 produced by these exponents.

Stifel then replaces multiplication of two numbers by addition of their exponents, and simplifies division to subtraction.

These four ideas contain principles behind the concept of logarithm. Are they enough to be called, together, the invention of logarithm?

Let's go on before we address that question.

If Stifel's idea is to be useful, one must enrich the two sequences of numbers so that multiplication of *all* numbers becomes possible.

Bürgi and Napier achieve this goal by different means.

Bürgi picks a base just a hair larger than 1.0, specifically 1.0001, and shows that with comparatively modest computational effort he can produce a table of logarithms with the desired feature.

As interpreted in the scaled table, the logarithms **0, 1, 2, ..., 23 027** of the list produce closely spaced numbers increasing from 1.0 to 10.0.

Napier invents a model of two moving points that result in the desired list of pairs of numbers.

As interpreted in the scaled table, he uses a factor less than 1, specifically $1/e = 0.3678\ldots$ The logarithms are for the sine values of angles from 0 to 90 degrees and thus range from infinity down to 0. But there are enough logarithms ranging from **2.302 5842** to **0** that cover the range 0.1–1.0 and thus support general computations.[151]

On to yet another idea, where the base of the logarithms is selected to be the same as the base of the underlying number system. With that idea, Briggs creates two lists that, in the scaled table of our interpretation, have a sufficient number of logarithms from **0.0** to **1.0** for the numbers ranging from 1.0 up to 10.0.

Finally, there is the gradual formulation of the logarithm function, which culminates in Euler's extension of the definition $\log_a(y) = x$ if $a^x = y$ to complex numbers by $\log_e(\cos(x) + i\sin(x)) = i \cdot x$.

Which of these six ideas should we declare to be the invention of the logarithm?

It may seem that the answer depends on what we consider to be the essence of that invention.[152]

- Is it Archimedes's notion that addition of exponents achieves multiplication of numbers?

- Or is it Apollonius's explicit use of exponents to define powers of the Greek number M, which had the value 10 000?

- Or is it Virasena's representation of any number by the product of a power of 2 and an odd number, and the related rules based on that representation?

- Or is it Stifel's arithmetic with two series of numbers that demonstrates how multiplication and division of particular numbers can be reduced to addition and subtraction, but cannot handle the case of general numbers?

- Or is it the construction of large tables by Bürgi and Napier so that generally multiplication, division, computation of powers, and extraction of roots are drastically simplified, but require the manipulation of complicating constants?

- Or is it the recognition of Briggs that, when the base of the logarithms is the same as the base of the underlying number system, such complicating constants can be avoided?

- Or is it Euler's extension of the logarithm function to complex numbers?

So, which of these steps created the essence of logarithm?

Stumped?

If you are baffled by the final question, you have in our opinion the correct intuitive response. The philosopher Ludwig Wittgenstein developed a methodology for analyzing such questions. When that technique is applied here, the question turns out to be fruitless.[153]

Indeed, we introduced the term "essence" and the related question only to demonstrate how abstract concepts such as "essence" can mislead us.

So let's abandon the fruitless search for the essence of logarithm and instead adopt a pragmatic approach where we determine how knowledge about logarithms grew. In a second step, we then decide who created the most important advance.

- There is no doubt that Archimedes's first use of powers of a number is groundbreaking. He did this in a world that only knew small numbers encoded by the letters of the Greek alphabet.

- The encoding of Archimedes's idea by Apollonius of Perga is then a natural formalization.

- Virasena's factoring of numbers involving powers of 2 results in new rules.

- Stifel's small table shows that in principle one may reduce computations with numbers represented by powers of a fixed base.

- Bürgi and Napier carry out a seemingly impossible task. They create large tables that achieved what would have looked like a dream to Stifel: simplified arithmetic for all numbers.

  By an ingenious choice of base, Bürgi creates the table in a few months, while Napier begins with a complicated model of movement of points and thus is forced to spend years on the construction of his table.

- Briggs improves the efficiency of the arithmetic by choosing the base 10. This requires a time-consuming construction he never fully completed.

102 WHO INVENTED LOGARITHMS?

- Lastly, there is Euler's expansion of the notion of logarithm far beyond arithmetic.

So which was the most important advance?

Stumped Again?

If you are baffled again, in our opinion you are right again, too: You are facing another fruitless question that cannot be answered.

So how do we respond if someone asks, "Who were the most important contributors to the idea of logarithm?" Or simpler, "Who invented logarithms?"

We need another model to address the question. How about deciding on the basis of what was accomplished relative to the knowledge and tools available at the time? And what the direct consequences of that work were?

As we go over the above summary for the various accomplishments, we see a spike of results relative to prior knowledge and of the impact on subsequent developments—the creation of the logarithm tables.

Here are arguments in favor of that evaluation.

At the time of Bürgi, Napier, and Briggs, the decimal number system has not only been adopted, but also been brought to the point where the notation is reasonably simple: A single marker separates the whole number from the decimal fraction.

Correspondingly, arithmetic can be performed manually with an effort that is small for addition and subtraction, quite substantial for multiplication, painful for division and computation of powers, and utterly frustrating for extraction of roots.

Into this world come the logarithm tables of Bürgi and Napier, and later that of Briggs. They reduce that effort by an incredible factor.

And thus, scientific exploration can proceed at a pace never imagined before. Indeed, an entire world of computing devices relying on logarithms rises. The development thrives for centuries, right up to 1976, the birthdate of the first electronic pocket calculator.

If you agree with these arguments, Bürgi and Napier become the candidates for being the inventors of the logarithm. Who did it first?

Deciding Who Was First

The criteria for resolving the question "Who developed idea X first?" are very stringent today. In an age where everything can be published with minimal effort, a claim of "I developed idea X independently, too" isn't accepted if the result was published a year ago by somebody else and could have been discovered with a reasonable search effort.

Put differently, the publication date generally determines priority. Yes, now and then we do allow for an exceptional evaluation that confirms the general rule.

But this criterion makes no sense when we go back several hundred years to the time of Bürgi and Napier. Communication across borders was complicated, and printing of large documents was costly and time-consuming.

Thus, inventions made years apart and without knowledge of competing efforts should still be considered independent, regardless when and how the results were published.

When we apply this widely accepted model to the case at hand, we must verify independence of the creative efforts of Bürgi and Napier, but can ignore precise timing of publication.

As for independence, this certainly applies to Napier, due to the unusual way he produces logarithm values via movements of two points. We know from Kepler's comment that Bürgi's construction precedes publication of Napier's work, so independence of Bürgi's effort is assured as well.

Thus, we conclude that Bürgi and Napier are independent co-inventors of the concept of logarithms.

We emphasize that this conclusion is based on the model involving incremental knowledge and impact on future developments.

If you agree with the use of that model, then you may agree with the above conclusion as well.

But if you have a different model in mind, you might examine how and why it differs from that employed here, and thus see why your final conclusion differs from ours.

The above reasoning may seem excessively detailed. We see next that there is good reason to delve into such minutiae.

20

Critical Comments

For the discussion to come, we display representative subsets of the scaled tables of Bürgi, Napier, and Briggs. Recall that the numbers in the left column of each table are black, while the logarithms in the right column are red, here indicated by bold font.

For clarity, we summarize the modifications that converted the original tables to the scaled ones.[154]

Bürgi's Scaled Table

We inserted decimal points into Bürgi's table.

y	$\log_{\text{Bürgi}}(y)$
1.000 000 00	**0**
1.000 100 00	**1**
. . .	
3.743 923 01	**13 202**
. . .	
9.999 997 79	**23 027**
. . .	
9.999 999 99	**23 027.0022**

Napier's Scaled Table

We inserted decimal points into Napier's table and used only numbers in the range 0.1 to 1.0 and their logarithms.

y	$\log_{\text{Napier}}(y)$
0.1 000 000	2.302 5842
0.100 188 1	2.300 7056
. . .	
0.510 543 0	0.672 2802
0.510 793 2	0.671 7905
. . .	
0.999 999 8	0.000 0002
1.000 000 0	0.000 0000

Briggs's Scaled Table

We removed redundant entries from Briggs's table, inserted decimal points, and sorted the entries.

y	$\log_{\text{Briggs}}(y)$
1.000 0	0.000 000 000 000 00
1.000 1	0.000 043 427 276 87
. . .	
2.000 0	0.301 029 995 663 98
2.001	0.301 247 088 636 21
. . .	
8.999	0.954 194 251 815 87
9.000 0	0.954 242 509 439 32
. . .	
9.999 9	0.999 995 657 033 47
10.000 0	1.000 000 000 000 00

We are ready to discuss arguments against our conclusion that Bürgi and Napier are independent co-inventors of logarithms.

There has never been any debate whether Napier invented logarithms. After all, he coined the word "logarithm"! Or whether he did so independently: His unique approach via two moving points clearly proves that.

The situation is different for Bürgi. Kepler's statement that Bürgi completed the tables many years prior to Napier's publication proves that Bürgi's work was independent and done prior to or at least concurrent with Napier's years' long effort. This proves independence.

But does Bürgi's table establish that he invented the concept of logarithm? We believe that our reconstruction of Bürgi's thought process proves this claim.

But us just saying so isn't good enough, since others have asserted the opposite. We now show that these objections rest upon a misunderstanding of Bürgi's work and/or application of inappropriate models.

There are two major objections. First, that Bürgi's table is just a table of antilogarithms. Therefore, the table doesn't prove that he invented logarithms.[155] Second, that his table contains some logarithm values, but that he didn't establish the logarithm as a mathematical function.[156]

We address these two objections.

Objection 1

The three scaled tables of Bürgi, Napier, and Briggs have one thing in common: Each is a collection of pairs of numbers. If you are given one number of a pair—be it black or red—then you get from the table the second number.

So for each black number, the table provides the red logarithm, and for each red logarithm, the table delivers the black number.

Saying that Bürgi's table is just a table of antilogarithms then means that for a given red number one can get the black number, but cannot do the reverse. This doesn't make sense.

It similarly wouldn't make sense to say Bürgi's table is just a table of logarithms. The table trivially has both features.

Remember our discussion in the introductory chapter? There we proposed that we would use the term "table of logarithms" as a matter of convenience.

No, no, somebody may object, you are overlooking something: A table of logarithms always has the feature that, in the case of decimal numbers, the difference between any two numbers is a constant decimal fraction. For example, we might have a table where the black numbers are $1\,000$, $1\,001$, $1\,002$, $1\,003$, ...

Of course, the logarithms of that table do not have such even spacing.

On the other hand—the objection continues—in an antilogarithm table the spacing of the logarithms is even, while the spacing of the numbers doesn't have this property.

So according to these two definitions—the argument goes—Bürgi produced a table of antilogarithms and not a table of logarithms.[157]

Can you see that this argument uses a model different from that employed by us?

So instead of discussing *conclusions* drawn from models, let's focus on the *choice* of models.

Often one may test appropriateness of a model by employing it in related settings and checking how well it works there.[158]

Here, Napier's table is a good test case. There is uniform agreement that it is a table of logarithms. But our scaled version of Napier's table, which handles general computations, doesn't have even spacing of the black numbers or the red numbers. Applying the above arguments, the scaled table then is neither a table of logarithms nor of antilogarithms.

No, no, somebody might argue, Napier's original table has evenly spaced angles—expressed in degrees and minutes—as black numbers, and then supplies the logarithm for the sine value of those angles.

Thus, Napier's original table is a logarithm table of the sine values of angles according to the definition requiring even spacing. But it isn't a logarithm table of numbers in general.

Let's see what Napier wrote about this.

The book *Mirifici Logarithmorum Canonis Description* contains the following highlighted statement.

Folio B of *Mirifici Logarithmorum Canonis Descriptio*, 1614.[159]

"The Description of the Wonderful Canon of Logarithms, and the use of which *not only in Trigonometry, so in all Mathematical Calculations*, most fully and easily explained in the most expeditious manner." [emphasis added][160]

In the Preface, Napier expands upon this claim.

funt obnoxię: Cœpi igitur animo revolvere, quâ arte certâ
& expeditâ, poffem dicta impedimenta amoliri. Multis fub-
inde in hunc finem perpenfis, nonnulla tandem inveni prçcla-
ra compendia , alibi fortaffe tractanda : verùm inter omnia
nullum hoc utilius , quod unà cum multiplicationibus , par-
titionibus, & radicum extractionibus arduis & prolixis, ipfos
etiam numeros multiplicandos, dividendos, & in radices re-
folvendos ab opere rejicit, & eorum loco alios fubftituit numeros, qui illorum munere fungantur per folas additiones,
fubftractiones, bipartitiones, & tripartitiones. Quod quidem
arcanum, cùm (ut cętera bona) fit, quo communius, eo me-
lius: in publicum mathematicorum ufum propalare libuit.

Preface of *Mirifici Logarithmorum Canonis Descriptio*, 1614.[161]

"For all the numbers associated with the multiplications, and divisions of numbers, and with the long arduous tasks of extracting square and cube roots are themselves rejected from the work, and in their place other numbers are substituted, which perform the tasks of these rejected by means of addition, subtraction, and division by two or three only." [emphasis added][162]

Surely Napier would have very much insisted that his table when restricted to general computations is also a table of logarithms.

Let's look at the effect of even spacing of black or red numbers. Regardless which case is at hand, the even spacing has no impact on computations except for interpolation. But interpolation is simplified by even spacing no matter which of the two types of numbers has such spacing.

Can you see now why we chose the term "table of logarithms" as described in the introduction?

If we had not done so and allowed evenness of spacing to control the terminology, then regardless of the choice we would have unfairly slighted Napier's work.

On the other hand, if Napier's table for general computation is to be a table of logarithms—in agreement with Napier's own choice of words—then it is only logical to apply the same term to Bürgi's table.

Indeed, using different terms would only obscure the common thread in Bürgi's and Napier's work.

Lastly, what might Kepler have said if somebody had claimed Bürgi's table to be just a table of antilogarithms that doesn't prove that Bürgi invented logarithms?

Kepler surely would have rejected that claim since his criticism of Bürgi's untimely publication of the table—discussed in Chapter 10—begins with the following sentence:[163]

"Such logistic numbers led Justus Byrgius to the *very same logarithms* many years prior to the appearance of Napier's system." [emphasis added]

Clearly, he had concluded that both Bürgi and Napier had created logarithms.

Let us proceed to the second major objection. It says that Bürgi's table doesn't represent a logarithm function.

Objection 2

The specific claim is, "Bürgi never had any notion of a logarithmic function, which is one reason why he cannot really be claimed as the inventor of the logarithms, even if his Progress Tabulen were made before Napier's tables.

"What Bürgi had was a correspondence between two progressions, one arithmetic, the other geometric, and a way to find a representative of a number within one of these progressions."[164]

The justification in part is based on the following statement of the author, which is declared to be a guess: ". . . Bürgi would have said that the red number of 360 is **128 099.78**, just as he said that the red number of 36 is [the identical] **128 099.78**." [bold font added in agreement with our color convention][165]

We saw during the discussion of Bürgi's instructions in Chapter 9 that this guess most assuredly is correct. Indeed, Bürgi declares[166] that 360 000 000 has red number **128 099̊789**, while 36 has **128 099̊ $\frac{78}{100}$**.

According to Bürgi's interpretation of the ̊, the first of the two red numbers corresponds to **128 099.789**, and the second one to **128 099.78**. The difference surely is due to an error in the original manuscript or introduced during copying.

The discussion of Bürgi's instructions in Chapter 9 included a fictitious interview where Bürgi explains his thinking when he constructed his table and the instructions for its use. Instead of just pointing to that imagined interview, we include here technical arguments that directly address the objection.

Let's look at a simple explanation for the assignment of the same red number to two numbers.

When looking up a red number for a given number, Bürgi just uses the sequence of the 9 most significant digits of the given number, padding with 0s if there are fewer than 9 digits. The decimal point of the given number is completely ignored.

There is an exception for the extraction of roots that we set aside for the moment.

This explanation is consistent with his conversion of final black numbers to the desired answers. That is, whenever the table supplies a 9 digit black number, the user needs to figure out the position of the decimal point.

This is evident when Bürgi writes at the end of the earlier discussed example[167] that the final 3 908 804 680 *"seindt die 9 ersten Ziffern des begehrten products"* (which are the 9 first digits of the desired product). That final number happens to be $3.908\,804\,680 \cdot 10^{17}$.

So the objection isn't that Bürgi fails to compute correctly, but that a red number is always selected on the basis of the 9 most significant digits of a given black number regardless of the placement of the decimal point. Thus, an infinite number of numbers are mapped onto each red number.

The logarithm function doesn't have that feature, since it is a one-to-one function. Hence, it may seem that Bürgi had no concept of the logarithm function.

In the same vein, one could criticize Briggs's original table and the scaled version derived from it here. The original table has only logarithms for the black numbers in the ranges 1–20 000 and 90 000–100 000, so it could be claimed that the table only partially represents the logarithm function of base 10.

The scaled version we have derived from Briggs's original table has only logarithms for black numbers in the range 1.0–10.0, so the same defect could be claimed.

Indeed, the very same complaint can be applied to modern logarithm tables, say of base 10: The decimal numbers and their logarithms are listed as sequences of digits, and the user needs to figure out the whole-number part of the logarithms.

Yet, nobody would deny that such a table is a discretized representation of the logarithm function. The representation just happens to be very compact and is based on an implied convention.

But more needs to be considered.

The function concept was introduced by Leonhard Euler (1707–1783) in 1755, more than 100 years after the time of Bürgi, Napier, and Briggs.

Thus, it isn't appropriate to criticize Bürgi for not describing his invention in the language of functions.

We should only check that the tables can be used in the intended fashion to carry out mathematical computations that implicitly rely on the full range of a discretized logarithm function. That surely is the case.

Indeed, if we allowed the criticism to stand that a mathematical development is recognized only if its formulation conforms with modern concepts, we wouldn't credit Leibniz and Newton with the invention of calculus.

With the function concept unknown at their time, they developed calculus to analyze the connection between variables and not—as viewed today—between variables and functions. Thus, an incredible milestone of mathematics would be redefined as just a modest step of mathematical progress.

———————————

In the final chapter we summarize the key results.

21

Conclusions

We have reached the end of our journey into the history of logarithms. Below are the main conclusions.

- Bürgi's table of logarithms is based on an ingenious construction that required just a few months of computation. In contrast, Napier's and Briggs's tables of logarithms were based on extensive computations that took years of effort.

- Based on the title page of Bürgi's table, it seems likely that he had the notion of a rudimentary circular slide rule. According to the evidence, he never pursued that idea. Most likely, he considered the accuracy of results to be inadequate. But if he had pursued that thought, he surely would have invented not just an elementary device relying on one or two disks, but would have come up with enhancements.

- Bürgi and Napier independently developed the concept of logarithm. Since Bürgi delayed publication of his table for at least 11 years, Napier for a long time was considered the sole inventor of logarithms. That evaluation is not correct. Most publications now agree that Bürgi and Napier are co-inventors.[168] But nagging doubts are still being raised. We have tried to refute such arguments by showing that they rest upon a misunderstanding of Bürgi's work and/or application of inappropriate models.

- Briggs's table of logarithms with base 10 is a significant advance since it completely avoids the use of annoying constants.

We close with a personal note.

For us, the investigation into the history of the logarithm has been full of surprises. We had always thought we fully understood the various developments: The concept supposedly followed directly from Descartes's notation of exponents for constants, and the construction of logarithm tables might have been tedious but nevertheless must have been rather routine.

But when we studied the history of mathematical notation some years ago, we became aware that Descartes defined his notation *after* logarithms had been invented, an unsettling insight. Indeed, we realized that we no longer had a reasonable explanation how the logarithm had come about.

That's when we started a detailed investigation, assisted greatly by several persons cited in the Acknowledgement section.

In the process, the idea emerged that we should place ourselves into Bürgi's time and position and see how he might have developed the logarithm idea using the tools of the day.

That process is very different from the way we carry out mathematical investigations today. For example, when we think about relationships connecting some variables, we immediately envision appropriate functions, often accompanied by graphs displaying key features. These tools didn't exist at the time of Bürgi, Napier, and Briggs.

And so we have come to greatly appreciate the seminal work of these mathematicians.

Notes

Throughout, "Wikipedia" refers to the English version.

Chapter 1 Introduction

1. [Rudman, 2007].

2. See Wikipedia "Slide Rule."

3. See Wikipedia "Difference engine."

4. See Wikipedia "Analytical Engine."

5. See Wikipedia "Ada Lovelace."

6. See Wikipedia "Konrad Zuse."

7. [Truemper, 2017].

8. [?].

Chapter 2 A Seemingly Simple Notation

9. See Wikipedia "René Descartes."

10. Source: https://en.wikipedia.org/wiki/Ren%C3%A9_Descarte s#/media/File:Frans_Hals_-_Portret_van_Ren%C3%A9_Descartes .jpg. "Frans Hals - Portret van René Descartes" after Frans Hals (1582/1583–1666) - André Hatala [e.a.] (1997) De eeuw van Rembrandt, Bruxelles: Crédit communal de Belgique, ISBN 2-908388-32-4. Licensed under Public Domain via Commons.

11. Source: https://en.wikipedia.org/wiki/Discourse_on_the_M ethod#/media/File:Descartes_Discours_de_la_Methode.jpg.

"Discours de la Méthode" by Unknown. Licensed under Public Domain via Commons.

12. It is convenient to allow m or n to be 0, by defining $a^0 = 1$ for any nonzero a. The case 0^0 is undefined.

13. See Wikipedia "John Napier" and "Henry Briggs."

14. [Staudacher, 2014].

15. [Waldvogel, 2014].

16. [Clark, 2015].

17. [Oechslin, 2001].

18. Interpolation is discussed in Chapter 6.

19. This aspect of Napier's table is discussed in detail in Chapter 20.

Chapter 3 Exponents

20. The entire chapter is based on [Cajori, 1928].

21. Source: https://en.wikipedia.org/wiki/Arithmetica#/media /File:Diophantus-cover.jpg. "Arithmetica by Diophantus." Licensed under Public Domain via Commons.

22. pp. 343–344 [Cajori, 1928].

23. See Wikipedia "Greek numerals."

24. See Wikipedia "Archimedes."

25. pp. 418–429 [Newman, 1956].

26. Archimedes by Domenico Fetti, 1620. Source: https://en.wik ipedia.org/wiki/File:Domenico-Fetti_Archimedes_1620.jpg. Painting of Alte Meister Museum, Dresden, Germany; see http:// archimedes2.mpiwg-berlin.mpg.de/archimedes_templates/popup .htm. Public Domain under US copyright code PD-old-100.

Chapter 4 Michael Stifel

27. See Wikipedia "Michael Stifel."

28. Source: https://en.wikipedia.org/wiki/Michael_Stifel#/me

dia/File:Michael_Stifel.jpeg. "Michael Stifel" by Unknown. Licensed under Public Domain via Commons.

29. Stifel, *Arithmetica Integra*, https://archive.org/details/bub_gb_fndPsRv08R0C/page/n5/mode/2up. Public Domain.

30. Source: "Acharya Virasena" photo by Samavasarana, own work, CC BY-SA 4.0, https://commons.wikimedia.org/w/index.php?cu rid=45730673. Photo reduced by K. Truemper.

31. For details about the Dhavala and Jain mathematics, see Wikipedia "Indian mathematics."

32. See Wikipedia "p-adic valuation."

Chapter 5 Jost Bürgi

33. See Wikipedia "Jost Bürgi." For details, see [Staudacher, 2014], [Clark, 2015], [Waldvogel, 2014], and [Oechslin, 2001].

34. Source: https://en.wikipedia.org/wiki/Jost_B%C3%BCrgi#/m edia/File:Jost_B%C3%BCrgi_Portr%C3%A4t.jpg. "Jost Bürgi Porträt" by User Dvoigt on de.wikipedia. Licensed under Public Domain via Commons.

35. Source: Proportional compass. https://de.wikipedia.org/w/i ndex.php?title=Datei:Buergi_zirkelgross.jpg&filetimestamp=2 0060923235538&. Public Domain according to Austrian, German, and Swiss copyright law.

36. [Oechslin, 2001] investigates in detail the mathematical results Bürgi must have developed for the design of the mechanized celestial globe depicted in this chapter.

37. pp. 189–197 [Staudacher, 2014] describes Bürgi's ingenious method for computing sine values and the construction of a sine table of extraordinary precision. The manuscript describing the method was considered lost until the late 20th century ([Folkerts, 2014]). There is an in-depth interpretation of the method using modern mathematics, see [Folkerts et al., 2015] and [Waldvogel, 2016].

38. Source: https://en.wikipedia.org/wiki/Jost_B%C3%BCrgi#/m edia/File:JostBurgi-MechanisedCelestialGlobe1594.jpg. "Jost Buergi-Mechanised Celestial Glob2." Photo by Horology - Own work. Licensed under CC BY-SA 3.0 via Commons.

39. Source: "Simon Stevin." Unknown author - Digitool Leiden University Library, http://socrates.leidenuniv.nl, https://commons.wikimedia.org/w/index.php?curid=72690. Public Domain.

40. See Wikipedia "Simon Stevin."

41. pp. 314–333 [Cajori, 1928] has details about the various contributors.

42. p. 314 [Cajori, 1928].

43. The full title is *Disme: the art of tenths, or decimall arithmetike teaching how to perform all computations whatsoever, by whole numbers without fractions, by the foure principles of common arithmeticke: namely addition, subtraction, multiplication, and division. Invented by the excellent mathematician, Simon Stevin. Published in English with wholesome additions by Robert Norton, Gent.*, London, 1608. Yes, there are two spellings "Arithmetike" and "Arithmeticke" in the title.

44. The notation is used in Bürgi's table of logarithms and the accompanying instructions. Elsewhere he also employed a subscripted "0", see p. 317 [Cajori, 1928] and p. 80 [Oechslin, 2001].

45. pp. 316–317 [Cajori, 1928].

46. p. 324 [Cajori, 1928].

47. [Roegel, 2010a].

Chapter 6 Bürgi's Construction

48. Interpolation: We consider here the simplest method, which is also employed by Bürgi. Let $(a, f(a))$ and $(b, f(b))$ be successive pairs in a table, where $a < b$ and $f()$ is a monotonically increasing function. Thus, $f(a) < f(b)$. Suppose we have a value x where $a < x < b$, and want to estimate the corresponding value $f(x)$. Since $f()$ is monotonically increasing, we know $f(a) < f(x) < f(b)$. We first compute the ratio $r = \frac{x-a}{b-a}$, and then estimate $f(x)$ by $z = f(a) + r \cdot (f(b) - f(a))$. This interpolation method is called linear because z is a linear function of r. In more complicated interpolation methods, not considered here, the estimating function for z is nonlinear.

49. pp. 98–99 [Oechslin, 2001] discusses the possible choices.

50. [Waldvogel, 2014] discusses how Bürgi most likely carried out checks for rounding errors.

51. [Waldvogel, 2014] includes a detailed discussion of the effect of interpolation of Bürgi's table entries on the precision of results. Here we try to recreate how Bürgi might have thought about that aspect.

52. Bürgi estimates 24.153 as logarithm value for 10.0 using $1.1^{24} = 9.849\,733$ and $1.1^{25} = 10.834\,706$ for interpolation:
$24 + (10.0 - 9.849733)/(10.834706 - 9.849733) = 24.1525... \approx 24.153$.
The exact value is $24.1588...$

53. Bürgi confirms the guess for the interval of x values from 0 to 1.0. The corresponding y values range from $1.1^0 = 1.0$ to $1.1^1 = 1.1$. For the evaluation of interpolation accuracy, Bürgi computes the correct y for several exponents x within that interval. He decides to divide the interval into eight evenly spaced parts with intermediate points $x = 0.125, 0.250, \ldots, 0.750, 0.875$.
Now $y = 1.1^{0.125} = \sqrt[8]{1.1}$, which he readily computes by taking the square root of the square root of the square root of 1.1, getting $y = 1.011\,985$. By repeated multiplication with that factor, he obtains the additional y values $1.024\,114$, $1.036\,388$, $1.048\,809$, $1.061\,388$, 1.074099, and $1.086\,972$.
He compares these numbers with the interpolated values 1.0125, 1.0250, 1.0375, 1.0500, 1.0675, 1.0750, and 1.0875. The corresponding differences are $0.000\,515$, $0.000\,886$, $0.001\,112$, $0.001\,191$, $0.001\,112$, $0.000\,901$, and $0.000\,527$. The maximum error occurs at the midpoint $x = 0.5$, just as conjectured.
We have carried out the computations in such detail to demonstrate that Bürgi indeed could confirm his guess with little computational effort.
Why wouldn't Bürgi simply plot the graph of the function to decide visually where the largest interpolation error occurs? The answer is simple. At Bürgi's time, such plotting had not been invented yet. Indeed, it requires the Cartesian coordinate system for the plane, which was invented in 1649, 17 years after Bürgi's death in 1632; see Wikipedia "Cartesian coordinate system."

54. Computation of interpolation error for an arbitrary interval: The difference $d(x)$ between the approximate and actual value is
$$d(x) = (1.1^{x+1} + 1.1^x)/2 - 1.1^{x+0.5}$$
$$= 1.1^x \cdot ((1.1+1)/2 - 1.1^{0.5})$$
$$= 1.1^x \cdot (1.05 - \sqrt{1.1})$$
Dividing $d(x)$ by 1.1^x, one obtains the relative error, which is the

desired $D_{1.1}$.

$$D_{1.1} = d(x)/1.1^x$$
$$= 1.1^x \cdot (1.05 - \sqrt{1.1})/1.1^x$$
$$= 1.05 - \sqrt{1.1}$$
$$= 0.001\,191.$$

55. Analogous to $D_{1.1}$,

$$D_b = (b+1)/2 - \sqrt{b}$$

Inserting the various values for b, the table values result.

56. The exact numbers are as follows:
$b = 1.01 : N = 231$
$b = 1.001 : N = 2\,303$
$b = 1.0001 : N = 23\,027$
$b = 1.000\,01 : N = 230\,259$

57. [Waldvogel, 2014] arrives at the same conclusion after a more precise analysis of interpolation errors.

58. [Waldvogel, 2014] discusses the computing process in detail, including controlling the buildup of roundoff errors by two additional digits of precision—called *guard digits*—and periodic checking for computational errors.

59. A complete analysis would have to consider the impact of inaccuracies introduced when logarithms are manipulated in additions, subtractions, and so on, and then rounded. The present discussion ignores these issues since Bürgi is just deciding which basis to use. For that purpose, the results described here suffice.

Chapter 7 Computation with Bürgi's Scaled Table

60. Source: Kepler, *Tabulae Rudolphinae*. `https://archive.org/details/tabulaerudolphin00kepl/page/2/mode/2up`. Public Domain.

61. Source: Briggs, *Arithmetica Logarithmica*, `https://archive.org/details/arithmeticalogar00brig/page/88/mode/2up`. Public Domain.

Chapter 8 Bürgi's Table of Logarithms

62. [Clark, 2015] has details about the copies of the table that have survived. The copy in best condition is in the Bavarian State Li-

brary (Bayerische Staatsbibliothek). It is accessible on the Internet: `https://daten.digitale-sammlungen.de/0008/bsb00082065/images/index.html?id=00082065&groesser=150%&fip=193.174.98.30&no=&seite=7`. [Clark, 2015] includes Bürgi's handwritten instructions, English translation, and commentary that covers the events surrounding Bürgi's creation of the table.

63. The entire table is available at `https://daten.digitale-sammlungen.de/0008/bsb00082065/images/index.html?id=00082065&groesser=150%&fip=193.174.98.30&no=&seite=8` of the Münchener Digitale Bibliothek. [Waldvogel, 2014] examines computing errors of the table in detail.

64. The entry 104080816 in the bottom right corner is not correct and obviously a typesetting error. The next page of the table begins with the correct value 104080869.

65. The interpretation is in agreement with [Waldvogel, 2014].

66. [Waldvogel, 2014].

67. pp. 24–26 [Roegel, 2010a] reviews various attempts to identify the base selected by Bürgi with concepts that did not exist at the time.

Chapter 9 Instructions for Bürgi's Table

68. See [Clark, 2015] and [Staudacher, 2014] about the history of Bürgi's handwritten instructions and the eventual printing.

69. [Gieswald, 1856]. [Clark, 2015] has the original text together with an English translation.

70. p. 27 [Gieswald, 1856]. [Gronau, 2016] has detailed information about the books by Jacob and Zons.

71. p. 29 [Gieswald, 1856].

72. [Gieswald, 1856]. The stated black number, 3 908 804 680, has 10 digits, and the rightmost 0 should have been omitted.

73. p. 31 [Gieswald, 1856].

74. p. 29 [Gieswald, 1856].

75. Source: `https://commons.wikimedia.org/wiki/File:Leonhard_Euler_2.jpg`. "Leonhard Euler 2" by Jakob Emanuel Handmann -

2011-12-22 (upload, according to EXIF data). Licensed under Public Domain via Commons.

76. Source: https://en.wikipedia.org/wiki/Function_(mathematics)#/media/File:Function_machine2.svg. "Function machine2" by Wvbailey (talk) - Own work (Original text: I created this work entirely by myself.). Licensed under Public Domain via Commons.

77. See Wikipedia "History of the function concept."

78. [Roegel, 2010a] has a detailed analysis of the various attempts to represent the table data by functions.

79. p. 99 [Oechslin, 2001] connects the terminating 0 of the red numbers with interpolation aspects.

80. Any logarithm function using the base 1.0001 and arbitrary scaling factors—see [Roegel, 2010a] for examples—will not support efficient computation unless modified by additional rules. This stems from the fact that direct use of the logarithm values would force addition or subtraction of unlimited multiples of a constant corresponding to the Bürgi constant. This even applies to multiplication and division, where Bürgi's process adds or subtracts at most the Bürgi constant.
See also the discussion near the end of this chapter. It connects Bürgi's use of significant digits for the selection of black numbers and the initial scaling step when Bürgi's scaled table is used.

81. You may want to use the English version of the instructions provided in [Clark, 2015].

Chapter 10 Bürgi's Title Page

82. Source: Title page. Toggenburger Museum, Lichtensteig, Switzerland. The museum kindly has granted permission to use the photo, which has been slightly improved for appearance.

83. We use again the convention that bold font stands for red color.

84. p. 230 [Staudacher, 2014].

85. p. 227 [Staudacher, 2014].

86. Source: https://en.wikipedia.org/wiki/Johannes_Kepler#/media/File:Johannes_Kepler_1610.jpg. "Johannes Kepler" by Unknown. Licensed under Public Domain via Commons.

87. [Kepler, 1627].

88. Source: Frontispiece. `https://archive.org/details/tabulaer udolphin00kepl/page/n1/mode/2up`. Public Domain.

89. Source: *Tabulae Rudolphinae*. `https://archive.org/details/ta bulaerudolphin00kepl/page/n37/mode/2up`. Public Domain.

90. The English translation of Kepler's comment is based on the German version of *Tabulae Rudolphinae*; see [Kepler, 1627].

Chapter 11 Geometric Computation

91. This simple observation has been made many times since the 19th century. We propose here that Bürgi himself was aware of this, and provide a conjecture why he didn't follow up on the idea.

92. Source: Title page. Toggenburger Museum, Lichtensteig, Switzerland. The museum kindly has granted permission to use the photo. Picture modified by K. Truemper.

93. Indeed, multiplication, division, and computation of low powers, say for exponents 2 or 3, can be done without the red numbers. But the extraction of roots requires the red numbers or—better yet simpler—a linear scale from 1 to 10 that we cover later.

94. Source: Ring of black numbers. Toggenburger Museum, Lichtensteig, Switzerland. The museum kindly has granted permission to use the title photo from which the black ring was extracted and modified by K. Truemper.

95. See Wikipedia "William Oughtred."

96. Source: `https://en.wikipedia.org/wiki/William_Oughtred#/ media/File:Wenceslas_Hollar_-_William_Oughtred.jpg`. "William Oughtred" by Wenceslaus Hollar - Artwork from University of Toronto Wenceslaus Hollar Digital Collection. Scanned by University of Toronto. High-resolution version extracted using custom tool by User:Dcoetzee. Licensed under Public Domain via Commons.

97. Source: Circular slide rule. Photo by Rod Lovett and Ted Hum of Oughtred Society. `http://osgalleries.org/classic/page2.cg i`. The Oughtred Society kindly granted permission to use the photo.

98. Source: Nested rings. Toggenburger Museum, Lichtensteig, Swit-

zerland. The museum kindly has granted permission to use the title photo from which the black ring was extracted and assembled to two nested rings by I. Truemper.

99. Slide rule with plastic disks using ring of black numbers cited above. By K. Truemper, photo released into Public Domain under Creative Commons CC0.

100. Source: Gunter's ruler. Photo by Rod Lovett and Ted Hum of Oughtred Society. http://osgalleries.org/classic/page2.cgi.
The Oughtred Society kindly granted permission to use the photo.

101. For a huge variety of slide rules, disks, and cylinders developed during the 350 years following Oughtred's inventions, see https://americanhistory.si.edu/collections/object-groups/slide-rules.

102. Thacher's slide rule. See Wikimedia https://commons.wikimedia.org/wiki/File:Senator_John_Heinz_History_Center_-_IMG_7824.JPG. Public Domain.

103. [Roegel, 2015].

104. Pocket-watch calculator KL-1. See Wikipedia "Slide Rule." Author Autopilot. https://en.wikipedia.org/wiki/Slide_rule#/media/File:Slide_rule_pocket_watch.jpg. License CC BY-SA 3.0 Unported. Instructions for the Russian-made calculator in English at https://www.sliderulemuseum.com/Manuals/KL-1_RussianCircularSlideRule.pdf.

Chapter 12 Design of a Circular Slide Rule

105. Essentially, the computation of powers and extraction of roots via a linear scale from 1.0 to 10.0 would simulate steps carried out with Briggs's table of logarithms discussed in Chapter 15.

Chapter 13 John Napier

106. [Havil, 2014]. See also Wikipedia "John Napier."

107. Source: https://en.wikipedia.org/wiki/John_Napier#/media/File:John_Napier.jpg. "John Napier" by Unknown - scanned from http://www-history.mcs.st-and.ac.uk/history/PictDisplay/Napier.html. Public Domain under US copyright code PD-old-

100.

108. Napier's Bones. Wikimedia. Author Stephencdickson. `https:` `//commons.wikimedia.org/wiki/File:An_18th_century_set_of_Nap` `ier%27s_Bones.JPG`. Processed to eliminate background. License Creative Commons BY-SA 4.0.

109. For a detailed discussion of Napier's Bones and various computing devices inspired by them, see `https://history-computer.c` `om/CalculatingTools/NapiersBones.html`.

110. For small dy, the time to go from distance y to $y + dy$ is approximately equal to $\frac{dy}{(N-y)}$. Using integration and denoting the natural logarithm by ln, we have

$$T(z) = \int_0^z \frac{dy}{N-y} = -\ln(N-y)\Big|_0^z = \ln(N) - \ln(N-z)$$

111. Source: Napier's *Mirifici*: `https://archive.org/details/miri` `ficilogarit00napi/page/n7/mode/1up`. Public Domain.

112. [Bruce, 2012].

113. The translation of the highlighted passage is taken from [Bruce, 2012].

114. Source: Preface of *Mirifici*. `https://archive.org/details/mi` `rificilogarit00napi/page/n11/mode/2up`. Public Domain.

115. Source: Partial page of *Mirifici*: `https://archive.org/detail` `s/mirificilogarit00napi/page/n83/mode/2up`. Public Domain.

116. [Bruce, 2012].

Chapter 14 Computation with Napier's Table

117. Source: First page of logarithm table of Napier's *Mirifici*: `http` `s://archive.org/details/mirificilogarit00napi/page/n73/mode/` `2up`. Public Domain.

118. Napier's logarithm 2.302 5842 is specified in LIB. I, CAP. IV, folio C3 recto, #9 of *Mirifici Logarithmorum Canonis Descriptio*. The value differs from the exact $\log_{1/e}(0.1) = 2.302\,585\,0929\ldots$ in the 6th position after the decimal point. Throughout we will use the former value to maintain complete agreement of the scaled table with Napier's values and instructions.

Chapter 15 Henry Briggs

119. [Bruce, 2004] contains biographical notes of Briggs. See also Wikipedia "Henry Briggs."

120. [Roegel, 2010c] reconstructs *Logarithmorum Chilias Prima*.

121. Source: Briggs's *Logarithmorum Chilias Prima*. http://www.pmon ta.com/tables/logarithmorum-chilias-prima/index.html. Also ht tps://commons.wikimedia.org/w/index.php?curid=43431356. Public Domain.

122. [Bruce, 2004] contains an annotated translation of *Arithmetica Logarithmica*. [Roegel, 2010b] reconstructs *Arithmetica Logarithmica*.

123. Source: Briggs's *Arithmetica Logarithmica*. https://archive.or g/details/arithmeticalogar00brig/page/n5. Public Domain.

124. Source: Briggs's *Arithmetica Logarithmica*. https://archive.or g/details/arithmeticalogar00brig/page/n129/mode/2up. Public Domain.

125. In the discussion below, we derive numbers and their logarithms directly from entries of Briggs's table. J. Waldvogel has pointed out that in several cases Briggs's table values are not quite correct. We ignore this aspect here to retain complete agreement of the scaled table with Briggs's table.

126. [Bruce, 2004].

Chapter 16 Comparison of Accuracy and Efficiency

127. [Waldvogel, 2014] thoroughly investigates the accuracy of computations done with Bürgi's table.

Chapter 17 Beyond Bürgi, Napier, and Briggs

128. [Craik, 2003] describes the life and achievements of Edward Sang, in particular the creation of the 47 volumes.

129. Source: "Edward Sang" School of Mathematics and Statistics, University of St. Andrews, Scotland. http://mathshistory.st-an drews.ac.uk/PictDisplay/Sang.html Public Domain under US copyright code PD-old-100.

130. See Wikipedia "Charles Babbage."

131. Source: "Charles Babbage" by Simon Harriyott from Uckfield, England - Charles Babbage. Uploaded by Oxyman, https://comm ons.wikimedia.org/w/index.php?curid=24731976. Also https:// en.wikipedia.org/wiki/Charles_Babbage#/media/File:Charles_Ba bbage_-_1860.jpg. CC BY 2.0.

132. See Wikipedia "Charles Babbage."

133. The two engines are not identical. One is the original design of difference engine no.2, and the other one, shown in the photo, is a later model. See Wikipedia "Charles Babbage."

134. Parts and Weight of difference engine: See http://www.comput erhistory.org/babbage/.

135. Source: https://en.wikipedia.org/wiki/Charles_Babbage#/m edia/File:Babbage_Difference_Engine.jpg. "Difference Engine No. 2" photo by User:geni. Licensed under CC BY-SA 2.0 via Common.

136. [Gronau, 2016] summarizes key events up to Euler's work.

137. See Wikipedia "Euler's formula."

138. We used this approach before as part of an investigation into the question whether mathematics is created or discovered, see [Truemper, 2017].

Chapter 18 Models of the World

139. [Hawking and Mlodinow, 2010]. See also Wikipedia "Model-dependent realism."

140. Source: Big Bang model by NASA/WMAP Science Team - Original version: NASA; modified by Cherkash. https://common s.wikimedia.org/w/index.php?curid=11885244. Public Domain.

141. Paragraph 6.36311 [Wittgenstein, 1963]. We use a direct translation of the German statement.

142. See Wikipedia "Human brain."

143. p. 211, 212 [Grafton, 2020].

144. Wikipedia "Checker shadow illusion." By Edward H. Adelson, own work, and vectorized by Pbroks13. https://en.wikipedia.o

rg/wiki/Checker_shadow_illusion#/media/File:Checker_shadow_i
llusion.svg. Licensed under CC BY-SA 4.0 via Commons.

145. Wikipedia "Checker shadow illusion." By Edward H. Adelson, own work, and vectorized by Pbroks13. https://en.wikipedia.o rg/wiki/Checker_shadow_illusion#/media/File:Grey_square_opti cal_illusion_proof2.svg. Licensed under CC BY-SA 4.0 via Commons.

Chapter 19 Who Invented Logarithms?

146. See Chapter 15.

147. See Chapter 10.

148. [Gieswald, 1856].

149. [Roegel, 2010a].

150. [Gronau, 2016].

151. The number **2.302 5842** is taken from Napier's instructions, see Chapter 14 for details. The exact number is **2.3025850929** . . .

152. One may be tempted to cast a wider net to capture early efforts involving the logarithm. In particular, one may claim that the positional notation for numbers uses logarithmic space to represent numbers, and thus is implicitly connected with the idea of logarithm.
We beg to differ. About 5,000 years ago, the Babylonians invented the positional numeral system with base 60. They left an open space when a position didn't have a nonzero. That caused confusion, of course, since one could misinterpret a number containing a blank position as two numbers. They later devised a symbol to represent this empty space.
Should we then declare that the Babylonians implicitly invented the idea of logarithm? In our opinion, we should not. If we do so, we apply terminology invented by Napier to the Babylonian system.
We may be tempted to take that step since we understand the profound connection of logarithms with many other ideas of mathematics.
But this doesn't imply that the Babylonians had any sense of logarithm where a value—the logarithm—is associated with each number in such way that rapid multiplication, division, extraction of roots, and computation of powers involving *any* numbers becomes

possible.

153. [Wittgenstein, 1958] provides a general methodology for investigating baffling philosophical questions. It uses the idea of *language games*. For a compact introduction, see [Fann, 2015]. Chapter 9 of [Truemper, 2017] contains a short summary. Below, we sketch arguments addressing the situation at hand.

The question "Which step created the essence of logarithm?" implicitly requires an answer to "What is the essence of logarithm?" For a resolution of the latter question, the following statement of [Wittgenstein, 1958] is helpful:

"116. When philosophers use a word 'knowledge,' 'being,' 'object,' 'I,' 'proposition,' 'name' and try to grasp the *essence* of the thing, one must always ask oneself: is the word ever actually used in this way in the language-game which is its original home?

"What *we* do is to bring words back from their metaphysical to their everyday use." [emphasis in original]

The arguments also apply when we examine an everyday word such as "table," "chair," or in our specific case, "logarithm," and try to define the essence in the abstract, so to speak—that is, when we have stripped away the actual settings of occurrences of the word and attempt to establish the essence disconnected from the varied uses.

Mind you, one can talk about the essence of something when restricted to a specific setting. But asking for the essence disconnected from the settings typically results in a confused philosophical discussion.

Chapter 20 Critical Comments

154. Recall that we have derived numbers and their logarithms directly from entries of the given tables. J. Waldvogel has pointed out that in several cases the table values are not quite correct. We ignore this aspect here to retain complete agreement of the scaled tables with the original ones.

155. The argument that Bürgi only produced a table of antilogarithms appears time and again in—sometimes subtle—statements denying that he produced logarithm values.

For example, pp. 216–217 [Grattan-Guinness, 1994] states, *"Bürgi did not speak of logarithms. He only applies the corresponding rules of calculating with powers and adjoined the 'red numbers' in an arith-*

metical series with double entry to the 'black numbers' in a geometrical series. From the method of construction this was later called an 'antilogarithm table', since the 'logarithms' (red numbers) are equally spaced whereas the Numeri (the black numbers) are variably spaced." [emphasis added]

156. [Roegel, 2010a].

157. Technically, the argument is not correct. At the end of Bürgi's scaled table, five logarithm values are very closely spaced so that the user has red values for black numbers near 10.0. The last of these red numbers is **23 027.0022**. It represents 10.0 with high precision. Thus, not *all* red values have constant spacing.

158. This idea is nothing but a particular version of Wittgenstein's notion of language games for the resolution of philosophical problems; see [Wittgenstein, 1958]. That is, each test of a model is a particular language game.

159. Source: Folio B of *Mirifici*: `https://archive.org/details/mirificilogarit00napi/page/n15/mode/2up`. Public Domain.

160. [Bruce, 2012].

161. Source: Preface of *Mirifici*: `https://archive.org/details/mirificilogarit00napi/page/n11/mode/2up`. Public Domain.

162. [Bruce, 2012].

163. See Chapter 10 for the complete statement in Kepler's *Tabulae Rudolphinae*.

164. [Roegel, 2010a].

165. [Roegel, 2010a].

166. p. 29 [Gieswald, 1856].

167. See Chapter 9 and p. 29 [Gieswald, 1856].

Chapter 21 Conclusions

168. For example, [Staudacher, 2014] argues emphatically for this conclusion.

Bibliography

[Bruce, 2004] Bruce, I. (2004). Briggs' ARITHMETICA LOGA-RITHMICA - translated and annotated. http://www.17centur ymaths.com/contents/albriggs.html.

[Bruce, 2012] Bruce, I. (2012). John Napier: Mirifici Logarithmo-rum Canonis Descriptio...& Constructio...- translated and an-notated. http://www.17centurymaths.com/contents/napierco ntents.html.

[Cajori, 1928] Cajori, F. (1928). *A History of Mathematical Notations, Vol. I: Notations in Elementary Mathematics.* Open Court Pub-lishing Company; go to https://archive.org/details/in.ern et.dli.2015.200372/mode/2up.

[Clark, 2015] Clark, K. (2015). *Jost Bürgi's Aritmetische und Ge-ometrische Progreß Tabulen (1620) – Edition and Commentary.* Birkhäuser.

[Craik, 2003] Craik, A. D. D. (2003). The logarithmic tables of Edward Sang and his daughters. *Historia Mathematica*, vol. 30; available at https://www.sciencedirect.com/science/arti cle/pii/S0315086002000186.

[Fann, 2015] Fann, K. T. (2015). *Wittgenstein's Conception of Philoso-phy.* Partridge Publishing.

[Folkerts, 2014] Folkerts, M. (2014). Eine bisher unbekannte Schrift von Jost Bürgi zur Trigonometrie. *Arithmetik, Geometrie und Al-gebra in der frühen Neuzeit.* Gebhardt, R. (Ed.), Adam-Ries-Bund, Annaberg-Buchholz, pp.107-114.

[Folkerts et al., 2015] Folkerts, M., Launert, D., and Thom, A. (2015). Jost bürgi's method for calculating sines. ArXiv.org:1510.03180.

[Gieswald, 1856] Gieswald, H. R. (1856). *Justus Byrg als Mathematiker, und dessen Einleitung in seine Logarithmen*. St. Johannisschule, Danzig, Prussia; available at Bayerische Staatsbibliothek `http://mdz-nbn-resolving.de/urn:nbn:de:bvb:12 -bsb10979407-8`.

[Grafton, 2020] Grafton, S. (2020). *Physical Intelligence: The Science of How the Body and the Mind Guide Each Other Through Life*. Penguin Random House.

[Grattan-Guinness, 1994] Grattan-Guinness, I. (1994). *Companion Encyclopedia of the History and Philosophy of the Mathematical Sciences*. Routledge Inc.

[Gronau, 2016] Gronau, D. (2016). Wie die Logarithmen zu ihren Namen kamen. Proceedings *XIII. Österreichisches Symposion zur Geschichte der Mathematik*, Österreichische Gesellschaft für Wissenschaftsgeschichte, Miesenbach, Austria, 2016.

[Havil, 2014] Havil, J. (2014). *John Napier – Life, Logarithms, and Legacy*. Princeton University Press.

[Hawking and Mlodinow, 2010] Hawking, S. and Mlodinow, L. (2010). *The Grand Design*. Bantam Books.

[Kepler, 1627] Kepler, J. (1627). *Tabulae Rudolphinae (Rudolphine Tables)*. `https://archive.org/details/tabulaerudolphin00ke pl/page/n1/mode/2up`. A book produced in 2014 contains the original Latin text and a German translation. It uses the fonts and graphics of the original book for both versions—an astonishing achievement. Title: *Die Rudolphinischen Tafeln*. Editor: Jürgen Reichert. Publisher: Königshausen & Neumann, 2014. See `https://www.amazon.de/Die-Rudolphinischen-Tafe ln-J%C3%BCrgen-Reichert/dp/3826053524`.

[Newman, 1956] Newman, J. R. (1956). *The World of Mathematics*, Vols. I-IV. Simon & Schuster; go to `https://archive.org/inde x.php` and search for "james newman world of mathematics".

[Oechslin, 2001] Oechslin, L. (2001). *Jost Bürgi*. Verlag Ineichen.

[Roegel, 2010a] Roegel, D. (2010a). Bürgi's Progress Tabulen (1620): logarithmic tables without logarithms. Research Report inria-00543936. https://hal.inria.fr/inria-00543936.

[Roegel, 2010b] Roegel, D. (2010b). A reconstruction of Briggs' Arithmetica logarihmica (1624). Research Report inria-00543939. https://hal.inria.fr/inria-00543939.

[Roegel, 2010c] Roegel, D. (2010c). A reconstruction of Briggs' Logarithmorum chilias prima (1617). Research Report inria-00543935. https://hal.inria.fr/inria-00543935.

[Roegel, 2015] Roegel, D. (2015). A new milestone: the first 7-8 places 2000 meters logarithmic slide cylinder. LORIA Research Report BP 239.

[Rudman, 2007] Rudman, P. S. (2007). *How Mathematics Happened*. Prometheus Books.

[Staudacher, 2014] Staudacher, F. (2014). *Jost Bürgi, Kepler und der Kaiser*. 4th edition. Verlag NZZ.

[Truemper, 2017] Truemper, K. (2017). *The Construction of Mathematics – The Human Mind's Greatest Achievement*. Leibniz Company.

[Waldvogel, 2014] Waldvogel, J. (2014). Jost Bürgi and the discovery of the logarithms. *Elemente der Mathematik*, vol. 69, pp. 89–117.

[Waldvogel, 2016] Waldvogel, J. (2016). Jost Bürgi's Artficium, an ingenious algorithm for calculating tables of the sine function. *Elemente der Mathematik*, vol. 71, pp. 89–99.

[Wittgenstein, 1958] Wittgenstein, L. (1958). *Philosophical Investigations*. Basil Blackwell; available at https://drive.google.com/file/d/0Bw-duXxYihdvWVlFaUhzclY5Vmc/edit.

[Wittgenstein, 1963] Wittgenstein, L. (1963). *Tractatus Logico-Philosophicus*. Routledge & Kegan Paul Ltd; go to people.umass.edu/klement/tlp/tlp.pdf for the German version and two translations into English.

Acknowledgements

The following persons and institutions provided information, contributed material, and greatly helped in the creation of the book:

B. Braunecker, R. G. De Cesaris, K. Clark, M. Grötschel, G. Gupta, L. Oechslin, M. Opperud, Oughtred Society, F. Staudacher, I. Truemper, U. Truemper, P. Ullrich, University of Texas at Dallas, and J. Waldvogel.

We very much thank these persons and institutions for their help.

K. T.

Index

accuracy, inherent
  Briggs, 83
  Bürgi, 25
  Napier, 83
analytical engine, 2
antilogarithm
  definition, 6
  spacing, 9
  table, 9
Archimedes, 13–15, 17, 18, 98, 100, 101
  picture, 14
ardha cheda, 18

Babbage, Charles, 87, 88
  analytical engine, 2
  difference engine, 2, 87
  picture, 87
base, definition, 6
Bernoulli, Jacob, 67
Briggs, Henry, 7–10, 34, 58, 75–85, 97, 99–102, 105–107, 113, 115, 116, 122, 126, 128
Bürgi, Jost, 7–10, 12, 13, 18–35, 37–54, 56, 57, 60–64, 66–68, 73–76, 78, 79, 81, 82, 84, 85, 97, 99–105, 107, 108, 110–113, 115, 116, 119–125, 128, 131, 132
  computing disk, 57
  constant (=23 027.0022), 28
  fictitious interview, 46
  instructions, 41

  picture, 19
  scaled table, 29
  scaling, 29
  table, 38

Cataldi, Pietro Antonio, 12
Chuquet, Nicolas, 12
circular slide rule, 2, 57
computation
  Briggs
    division, 80
    multiplication, 80
    powers, 80
    roots, 80
  Bürgi
    division, 32
    interpolation, 34
    multiplication, 31
    powers, 32
    roots, 33
  interpolation, 23, 84
  Napier
    division, 75
    multiplication, 75
    powers, 75
    roots, 75
cylinder slide rule, 2

decimal
  notation, 20
    Bürgi, 21

Napier, 21
point, 20
system, 20
Descartes, René, 5, 6, 14, 21, 23, 30, 39,
    43, 45, 46, 48, 49, 78, 98, 116
picture, 5
Dhavala (commentary), 18
difference engine, 2, 87
    no. 2, 87

Euler, Leonhard, 45, 46, 88, 89, 97, 100,
    102, 113, 129
logarithm function, 88
picture, 45
exponent, definition, 6
    by Stifel, 16

Gieswald, H., 41
Gunter, Edmund, 7, 58, 59, 76
line, 58
ruler, 58

Hawking, Stephen, 90

instructions (Bürgi), 41
interpolation, 23, 82

Jacob, Simon, 41, 123

Kepler, Johannes, 13, 19, 21, 34, 52, 53,
    97, 103, 107, 111, 122, 125, 132
picture, 52
Tabulae Rudolphinae, 53

Leibniz, Gottfried Wilhelm, 14, 67, 113
logarithm
    definition, 6
    function, 88
    spacing, 9
    table, 8

Mlodinow, Leonhard, 90
model-dependent realism, 90
    errors, 93

facts, 91
true facts, 95
myriad (number M), 13

Napier, John, 7–10, 21, 52, 53, 64–79, 81–
    84, 97, 99–111, 113, 115, 116, 118,
    127, 130
Bones, 65
constant, 74
picture, 65
scaled table, 73
scaling, 74
table, 67
Newton, Isaac, 14, 67, 113
Norton, Robert, 21
number
    black (Bürgi), 38
        interpretation, 44
    Greek letters, 13
    guard digit, 122
    red (Bürgi), 38
        interpretation, 37, 38, 48

Oughtred, William, 7, 57–59, 126
circular slide rule, 57
picture, 57
slide rule, 59

power, definition, 6
progression
    arithmetic, 16
    geometric, 17

van Roomen, Adrian, 12

Sang
    Edward, 86
    Flora Chalmers, 86
    Jane Nicol, 86
scaling
    Briggs, 78
    Bürgi, 29
    initial
        Briggs, 80

Bürgi, 29
  Napier, 74
  Napier, 74
slide rule, 2, 59
  additional scales, 62
  circular, 2, 57
  construction effort, 62
  cylinder, 2, 59
  largest, 59
  pocket watch KL-1, 60
  spacing of ticks, 61
spacing
  antilogarithm table, 9
  logarithm table, 9
Stevin, Simon, 20, 21
  picture, 20
Stifel, Michael, 16–18, 41, 42, 99–101, 119
  picture, 16

table
  antilogarithms, 9, 10
  construction vs. use, 10
  logarithms, 8, 10

Briggs, 77
Bürgi, 38
Napier, 67
scaled
  Briggs, 79
  Bürgi, 29, 30
  Napier, 73
Tabulae Rudolphinae, 53
Thacher's slide rule, 59
title page of table
  Briggs, 76
  Bürgi, 50
    nested rings, 57
    ring of black numbers, 56
    sweeping angles, 55
  Napier, 67

Virasena, Acharya, 18, 99–101
  picture, 18

Wittgenstein, Ludwig, 91, 101, 132

Zons, Mauritius, 41, 123

Milton Keynes UK
Ingram Content Group UK Ltd.
UKHW041453041123
431960UK00001B/92

9 780999 140208